Looking for The Way

To Linda

who taught me so much

Looking for
The Way

Two systems of a small planet and
how the Universe works

Jeremy Flint

Moorcroft Books

Copyright © Jeremy Flint 2014
First published in 2014 by Moorcroft Books
New Brighton, Mold, Flintshire, CH7 6RF

Photo images *copyright* Maddocks / Flint

Graphics by No Duff Stuff
www.noduffstuff.co.uk
(based on designs by Jeremy Flint)
Distributed by Lightning Source

The right of Jeremy Flint to be identified as the author of the work has been asserted herein in accordance with the Copyright, Designs and Patents Act 1988.

All rights reserved. This book is sold subject to the condition that it shall not, by way of trade or otherwise, be lent, resold, hired out or otherwise circulated without the publisher's prior consent in any form of binding or cover other than that in which it is published and without a similar condition including this condition being imposed on the subsequent purchaser.

ISBN 978-0-9928451-0-0

3 5 7 9 10 8
6 4

This book production has been managed by Amolibros
www.amolibros.com
Printed and bound by Lightning Source

Contents

A Seed is Sown		*page 6*
Branch 1	*The Journey Begins*	*page 9*
Branch 2	*The Working Day*	*page 20*
Branch 3	*Early Learning*	*page 29*
Branch 4	*Pyramids at Work*	*page 37*
Branch 5	*The Birth of Curiosity*	*page 51*
Branch 6	*Job Analysis*	*page 60*
Branch 7	*Autumn Advances*	*page 70*
Branch 8	*First Birthday*	*page 82*
Branch 9	*The Blizzard*	*page 98*
Branch 10	*Galum's first Christmas*	*page 115*
Branch 11	*Stock-Taking*	*page 132*
Branch 12	*Enter Joe Public*	*page 156*
Branch 13	*Linda's Game*	*page 174*
Branch 14	*Oak Lightning*	*page 187*
Branch 15	*Greek Excursion*	*page 200*
Branch 16	*Tree Counting*	*page 220*
Branch 17	*Loose Ends*	*page 228*
Branch 18	*The Way*	*page 241*
Branch 19	*Purpose*	*page 261*
Epilogue		*page 280*

A Seed is Sown

If time does exist, as most people believe, then it really does fly.

It is hard to imagine that this little planet, upon which I find myself, has circled the sun thirty times since I first tripped over the fundamental law of the Universe.

All those years ago, in the mid nineteen-eighties, I was simply looking for the way. By this I mean I was thinking about what was the real purpose of living. What did it all mean? Was there a goal and if so, how best to achieve that objective?

It struck me that a lifetime could be wasted pursuing things that really did not matter (in the long run) unless these questions were considered near the beginning. Daily actions and events may have made some cultural sense at the time, but then again, they may have all led to the wrong goal.

As these thoughts developed I was unsure if they were right or wrong. Hence I decided to simply record them and see how they evolved.

So, thirty planetary revolutions later, I embark upon an attempt to explain my findings. I feel I will fail from the outset, as most people have no interest in the overall puzzle in which they are embedded; they are simply too busy trying to make it through life's journey. The overall picture is hard to see with so much noise drowning out any perceivable signal. Sometimes there isn't time to sit around and wonder why the wheel is turning when you simply have to hold on tight just to get by.

It may have been better if I had been a trained philosopher, or physicist, but I had been trained in other areas: I had been trained to make things work; specifically, electrical things, so these options were beyond me. Not gifted in logic nor mathematics, I had to use my artisan skills to make a living. In my limited spare time I read factual books, reference works, scientific papers and magazines and, as the years rotated, I became good at recognising systems, the way they worked, their interactions and their modes of failure. Eventually I worked my way up the career path, though at meetings I introduced myself as a "time-served electrician working my way down" as I often felt my career progress always took me further away from the truth I was seeking. Nevertheless, my opening phrase always broke the ice.

On my lapel badge was "Experimental Specialist", about as high as a technician could get without turning to the dark side of administration or management. I always preferred science and the laws of physics; projects and people came and went but the laws of physics remained intact, off-limits to change by administration and management, who would have happily changed them if they could! But they seemed content to rearrange our chains of command and working structures ad infinitum. Meanwhile, I evolved from one project to the next, making things work in accordance with the established laws of physics and engineering. So most of my story is set in an apparently bygone time, though there are remarkable similarities with today, even though three decades have elapsed.

To distinguish between my present thoughts (ten years into the new millennium) and that bygone age, I use bold type (and sometimes colour) for the modern writing **and** normal typeface and black and white, or sepia tint, for the earlier composition and images.

In this manner we now return to the start of my musings in the latter half of the twentieth century, back to a simpler yet busier time for me. I wasn't sure where to begin, yet now I don't think it really matters. As long as the focus is on the goal, then there are many different ways to get there. This book is just the route that I took.

Branch 1

The Journey Begins

The alarm clock rings; well, not so much a ring, more a clattering, clanking sound. First observation: nothing is perfect.

It's six-thirty. I've been semiconscious for half an hour, disturbed by the sunrise lighting up the bedroom. I make a real effort to fill my lungs with cool air, ready to face another day. Deep breaths; I roll out of bed.

"I'm going," I say as a matter of routine to my wife, Tracy. No reply today.

I dress, and then enter the kitchen. Fill kettle, switch on, teabag in teapot. Pan of milk on front ring. Switch to "LO", for a wash I go. Obviously, I'm still on automatic pilot; no conscious thought involved yet. Wash, shave, clean teeth, comb hair and return to the kitchen. Milk in cups: sugar in right-hand cup. Warm milk on the cornflakes (which were put out last night) and switch oven off.

I deliver the tea-without-sugar to Tracy and return to the kitchen once more to eat my breakfast.

Jacket on, keys in pocket? Yes. Money? Adequate. One final important task left. I look in to see if my seven-month-old daughter Linda is safe and sound. She is usually restless at this time of day, especially if the Sun is illuminating her window, too. Daylight is her second timekeeper, her tummy being the first; so on bright mornings she wants breakfast earlier than on duller days.

I enter her room and she turns to look at me over her shoulder.

"Hello, Linda," I whisper. "It's only me, Dada."

She smiles and then her head flops against the safety mattress again.

"I'm going now ... to work." She doesn't understand a word I say, but the intonation of my voice is friendly, playful. "We'll have a good play tonight," I promise. One day soon I'll have to be careful what I say. Some nights I have to cut lawns, mend the car, fix a vacuum cleaner, or do a host of other jobs too diverse to mention.

"I'm going now, Tracy. See you tonight."

"See you," she replies.

I have to walk one mile from my bungalow to the bus stop; there I catch a private bus which takes me to a factory. It's a pleasant walk along a busy lane joining two main roads. The journey can be divided into two. The first half has open fields on either side stretching out as far as the eye can see to a distant hedgerow on the left, or a valley with hills beyond on the right; this is the uphill part of the journey. On the downward slope bungalows and houses border the road, giving it a suburban landscape with plenty of variety – more than usual, I would say.

The gentle uphill walk can be really invigorating, especially on a lovely sunny morning. Sometimes I actually feel *privileged* to be up so early to witness this panorama while it is still peaceful and calm; a sharp contrast to the initial feeling when the alarm clock first clanks.

The footpath is on the left as I ascend the hill and running parallel to it is a neatly trimmed hawthorn hedge. The elderly farmer, who lives at the apex of the hill, clips the hedge twice a year. He's still cutting it with hand shears and this creates a random undulating wave along the top of the hedgerow, which, for reasons as yet unclear, tends to stand out. Perhaps I have been working with geometric regularity for too long: installing vertical or horizontal conduits, never curvy – well, only once to blend into a bent breeze block wall. Now something as sporadic as a wavy hedgerow looks out of place. An occupational hazard, I guess.

There are two yards of wild grass verge between the path and the corrugated hedgerow and it is here that I make my second observation of the day. What we are about to encounter is so familiar that it has become an accepted part of our existence. It comes in a variety of colourful textures and forms. Aha! Here is the first example now and a classic one, too: the non-returnable, throwaway, crushed Coke can in gaily coloured red and white.

Rubbish. One of man's less endearing behavioural characteristics is to take nature and mottle it with rubbish. It is everywhere. Each stride now brings into view more waste material. A lollipop stick, crisp packet, black plastic dustbin liner, cigarette stubs, an old newspaper, a non-returnable glass bottle, aluminium foil – Chinese take-away style – and a plastic cup.

Most of this assortment is blown off the path by the prevailing westerly wind, and then it is either caught by the spiky hawthorn hedgerow, or ends up in the field for the cattle to chew on. But this does not mean the path is clear, far from it. The path is dotted with dog poo; dog droppings all over the place. In dry spells the path becomes an uphill slalom course; left through the brown bits, right through the fawn ones – horrible thought!

Farm buildings stand at the top of the hill, dwarfed by high voltage power cables (400kV – super grid) suspended on flimsy-looking pylons. When the weather is wet, or even just moist, the cables crackle and hiss as if in protest. I found this phenomenon quite disconcerting at first and made a point of walking underneath them quickly until out of earshot. Silly, I suppose, but today the Sun has lulled this powerful energy web into a gentle slumbering buzz.

Just beyond the farm is the infant school where Linda will one day begin her formal education: a slow methodical enlightenment towards adulthood. I sometimes study her when she is playing, examining an empty yoghurt tub or a squeaky toy. She has got the whole world to learn. It seems incredible that her tiny scalp contains the potential to fulfil this task. It is a tremendous challenge for each individual.

I imagine spending five years teaching Linda how to be a good girl, then ten years trying to explain why the rest of the world isn't so good. We will see. In the meantime I had better start working out some answers to life's big questions.

At the halfway stage of the journey I look down the hill onto rooftops below. The second half of the route can be divided longitudinally, so that on the left are bungalows and houses, whilst on the right is the public highway. Between the two is a dark tarmac path and in the middle of it, along its entire length, the telephone people have dug a trench. After burying cables, the trench has been capped with light grey gravel, which stands out like a sore thumb. I make a habit of walking along the trench top, because it represents, in my mind, a kind of no-man's-land. To the left of the dividing line is private property; to the right a public road. The trench-top is a visible interface between the two.

If we catalogued all the rubbish seen earlier, we might expect the creatures that created such a wealth of waste would surely be living in squalid surroundings. Their dwellings must be brimming with junk and the gardens swamped with the overflow. Yet an examination of their properties reveals quite the reverse.

There are some beautiful gardens on my left. Hours of thought and labour and expense have been lavished upon them. Lawns like green velvet surrounded by gaily coloured borders, then shrubs and finally small brick walls marking the outer boundaries. Most of the gardens are good, but some are superb. I am passing one of the best now. The lawn, on close inspection, is not as good as others, but the eye is drawn away from this by a beautifully simple and carefully blended border. There is blue lobelia, orange tagetes, white alyssum, golden marigolds and mixed antirrhinum. They have been spaced so that now, in full bloom, the mosaic is complete without looking crowded.

In contrast to this tapestry of colour, dwarf conifers, in subtle hues of silver and green, surmount a small rockery, providing a pastel backcloth. The aubrietia and golden alyssum at the rockery base have faded now; they had their flourish earlier in the year.

There is another quality garden farther along. The lawn on this occasion is the centrepiece. It is smooth and as verdant as a snooker table; a green even-textured baize. Around the lawn are numerous roses. The drive gatepost has been partly overgrown by a healthy blue hydrangea, while a symmetrically trimmed pine hedge separates this garden from the next door neighbour.

So contrary to what one might expect, there is no rubbish in most of the gardens. Instead it is all strewn out here to the right of no-man's-land: there is no rubbish on the private property, it is here, in the gutter and on the road in the public domain.

There is something strange here that puzzles me. How can people who adore order and tidiness in their gardens, suddenly and totally neglect it two yards away across the footpath? What kind of thinking underlies this attitude?

Perhaps most of the litter is dropped by children. There were a lot of lollipop sticks and sweet wrappers among my survey. Yet these children are the sons and daughters of the people who own the beautiful gardens. Surely they are told not to litter the streets?

It appears people feel no responsibility for those areas that do not directly belong to them. I remember a squashed, desiccated hedgehog lying by the gate of a bungalow, not quite within an invisible line marking out the private property. It lay there for weeks. I have seen a milk bottle in the gutter outside the drive of a house. A week later it was broken; then it gradually fragmented and disappeared by reduction. People seem to have no desire to keep tidy those areas bordering their dwellings. I wonder why they no longer care.

I reach my destination, one of the most public places one can imagine: a bus shelter; a small breeze-block oblong, ten feet by four by seven feet high. The roof is corrugated asbestos and the front is open. Inside, the cement base is littered with waste: beer cans (buckled and crushed by frustrated energy), newspapers and greasy chip paper, and chips as well! Cigarette stubs and matches radiate out from the front of the shelter towards the kerb, as if they have been scattered like seeds on the land.

I usually wait outside the shelter. That way it is easier to flag down the bus and it is healthier, too.

To become a qualified bus traveller requires years of hard training. One's mind has to be elevated to the level expected of a Tibetan monk, though some passengers prefer sleeping. Either way, you have to be in a trance-like state, then, while your body is subjected to extremes of discomfort, you may attain nirvana – total oblivion of the self. This helps reduce one's blood pressure when the bus is late. If I recall the worst journey in my experience, it may further your knowledge of commuting.

Picture the scene: a bleak winter's morning by the roadside awaiting the arrival of the B37 bus. As usual, I am the first to arrive. The time is ten minutes past seven, the bus is due at a quarter past, though rarely arrives until twenty-five past. Everybody knows this and the rest usually arrive at twenty past. (Hope that's not too complicated.)

It came on time once – only I caught it that day.

It is cold. Even I can feel it inside my specially designed duffel coat. I must explain at this point that Tracy is a needlework teacher and can produce any clothing required cheaper and better than you generally buy in the shops. The coat is calf-length, lined with a smart woollen tartan and an interlining of nylon to make it impervious to wind and light rain.

There was a heavy frost last night and the white frozen grass indicates it is still below freezing point. Today I stand inside the shelter, ankle-deep in chip paper and look bleakly across the road at a black Alsatian dog sniffing the base of a lamp post. Buses do not like this weather; who does? But some buses don't like this weather *so much* they won't come out in it!

A few colleagues arrive and we exchange *un*pleasantries about the temperature. There is some mention of a brass monkey looking for a welder, which I don't understand, but we all laugh anyway and our breath forms a cloud of white vapour which limply drifts out of the shelter above our heads. Some wit, wearing a fashionable bomber jacket, asks if they make duffel coats like mine in long sizes. I reply that at least my brain is warm, whereas his nether regions are severely exposed. He hops from foot to foot, doing a little jig to keep warm. I stand still; warm yet unfashionable, it seems.

No sign of the bus yet.

Two more fellow travellers arrive. There is limited chip-free standing room for five people inside the shelter, so the latest arrivals are left hopping about on the pavement, subjected to the ram air from passing traffic.

It is half past seven now and an air of gloom is hanging over our little group. This is where one's training becomes vital. Who will be the first to crack up, develop a bad back and return to a warm bed to try again tomorrow?

Unconsciously, we huddle together, like penguins on an ice floe. In fact, this analogy is a very accurate description of the scene in front of me now. Our happy little band (or gaggle) is waddling from foot to foot, all relentlessly staring down the hill, waiting, willing the bus to appear. Some are of the opinion that spring might arrive first.

Three buses have already passed by; here comes the fourth. We make a tentative waddle towards the kerb. No, it is not our bus. We wave it on and stare daggers at the innocent driver.

Eventually our bus is sighted and with mixed feelings of relief and dismay we ascend the steps, pay our fare and select a seat. Ooh! Cold, simulated leather seats! These buses must be designed by warm-blooded masochists. Your feet are freezing and where do they put the heater outlets? In the ruddy roof, that's where! It is a good job my duffel coat offers some protection; but imagine what the fashionable must be feeling now?

Our driver offers an explanation. Another driver didn't turn up for work and then the bus wouldn't start. There is no response; we are all too numb.

Well, I suppose these things can happen and remember this is the worst journey I can recall. But what puzzles me is how a professional bus company can get itself into such a poor state of organisation. If the forecast says the weather is going to be bad most people would prepare for it. The professionals apparently do not. Come rain or shine, it is all ad lib at the depot.

When snow is forecast I bring my wellies in from the garage and set the alarm clock to wake me five minutes earlier. That is all I need to compensate for the harder trudge over the hill. It is annoying when you make the effort to walk a mile and arrive on time, then are subjected to the almost arbitrary arrangements of an undedicated bus company.

After six miles of the worst journey I can recall, the bus breaks down. Later a maintenance crew appears, disappearing under the front of the bus to unfreeze the compressed air brakes with a blow torch. Well, at least the bus will be warm. It might feel better when its compressor nuts are roasted and then agree to take us to work.

I must confess a degree of sympathy for the bus drivers who take the brunt of public anger on such occasions. Each one responds differently with a characteristic individuality that is stamped on their driving. Regular travellers have even suggested nicknames for some of the more eccentric drivers: "L'escargot" (the snail) is the first to spring to mind. He's an elderly gent, very polite, handles every bus as if the engine is being run-in, habitually allowing everybody to pull out from side roads and roundabouts. This altruism catches people off guard and consequently their cars often jump forward and stall, as politeness from a bus driver has been known to cause cases of shock. So now they are stuck halfway across the road: but it doesn't matter, we will wait. We are in no hurry. Restart the engine, get it back in gear. Fit an exchange gearbox if you like. Want a push? There is no rush – we're already late.

The snail does have one point in his favour, a remarkably smooth gear change. In fact, it is so fluent that you are unaware of the event. So when the snail drives, the experienced passenger takes advantage of the circumstances and settles down for a long snooze.

Another driver has been dubbed "the most relaxed bus driver in the world ever". He has earned this distinction by consistently failing to bother looking at the road. Granted, he does occasionally have a sideways glance at the traffic, but his main preoccupation seems to be at right angles to the direction of travel. Hedgerows, trees, a windmill, they are all more intriguing than the control column of a bus.

Meanwhile, we transfer from our frozen bus, with roasted compressor nuts, to a working model and continue out of the town, along lanes, past hedgerows, trees and a windmill and I secretly wonder if mankind progressed from the windmill age a little too soon. Perhaps the cold is getting to my mind.

This story is not directly about the private versus public dichotomy; differing opinions will always exist. Someone who actively believes in one ideal automatically generates a reaction. Thesis and antithesis are born at the same instant, a mischievous arrangement which throws everything open to debate. Differing points of view rarely converge; on the contrary, a hardening of ideas can cause a problem to split into two radically opposed solutions, while the answer is usually lying quietly in the middle. I want to consider a few of the more common discrepancies, examine them and perhaps uncover a fundamental reason for their existence. Only a fool would try to resolve these matters. I only wish to understand them.

 The bus arrives late.

With the passage of time I can contentedly report that local mankind reacted to dog poo with signs, one hundred pound fines, and also poop bins for the owners to scoop and bag the droppings into. Now the footpaths are largely clear of this hazard.

The old bus stop didn't survive though, and has been replaced with a modern open-plan frame with a neat digital display signalling when the next bus is expected to arrive. The old bus company went out of business and a myriad of new companies filled the gap.

The works' bus was stopped as a cost-cutting exercise. I travelled on it for eighteen years. Then I and about a dozen fellow travellers had to buy cars and join the rush-hour traffic. Ironically, this was at a time when the factory public relations people were proclaiming their green agenda, whilst putting more cars on the road.

All the litter I encountered is now subject to recycling. Back in days gone by we only had one paper bag for household waste into which everything went, later to be buried in a land-fill site. Now there are several coloured bins and bags, which take up a lot more space, where litter is pre-sorted for collection. If there is windy weather on recycling day I make it my duty to gather the blown debris from my adjacent road area. Litter still belongs to nobody when it is on the streets.

Remember, this is a time before the advent of the PC – the now ubiquitous Personal Computer. My original manuscript was handwritten and only later printed out on a dot-matrix printer. Office programmes for spreadsheets and drawing packages were still in the future. Easy inclusion of an image into a document was unheard of at this time as the digital camera had not been invented. The World Wide Web was not in the mainstream and in offices throughout the land typing pools and purchasing departments still worked on paper documents by hand.

Oh, how all this was to change!

Branch 2

The Working Day

I am an electrician. I earn my living by installing and maintaining things of an electrical nature − that is fairly obvious. I work in a typical factory and it was here I encountered a personal conflict which I resolved with the ideas forming part of the basis of this dissertation. I found the answer to my problem had potential applications to other areas of discontent, more of which we will encounter later.

A factory can be considered as a microcosm of society. Within the organisational framework, specialist groups operate. Each has to marry independent action with the need to compromise and achieve common objectives. Hence, a factory is an excellent place to study conflict.

Electrician: I find it a fascinating trade (that is not so obvious) and sometimes people ask me if I could "do better" − a tricky question. Yet, even humble electricians can have ambition. Once I wanted to be the best electrician in the world and then found local conditions unsuitable. The environment is a big factor; I mean, it is unlikely the best electrician in the world is employed behind the light bulb counter at some superstore. He just would not have the opportunity to see all the various applications of electrics, even though he may be a leading authority on light bulbs. No, the best electrician in the world is not a specialist. Instead he is well versed in the knowledge of anything in which electrons move. It is the variety that creates the interest and the interest that brings out the best in him. He could tell you the cause of a thunderstorm, or fault-find on a complex electronic assembly. He doesn't do this with a phenomenal memory or an IQ of 138, but by learning basic procedures that break problems down into solvable bits. In short, he's an analyst. The techniques he employs are not taught in colleges or universities: there they seem more concerned with the objectivity of electrics: naming items and defining their individual purposes. His knowledge is an empirical framework built upon years of observation. He knows what the component parts are called and is now more interested in the system and the events occurring in systems: dynamic, not static electrics. It is a conscious choice between aptitude and theory; another dichotomy.

When a problem occurs on a process line the electrical department is frequently the first to be called out; but, on examination, the fault may prove to be a broken coupling (mechanical) or an instrumentation error. Now the electrician has to decide which category, box or pigeonhole the failure fits into and the correct action to take; in other words, whom to notify. So he has to understand another kind of system, too, the organisational hierarchy governing his work, a subject rarely covered at all in the classroom.

But all this does not make him a genius, far from it. He even struggles with simple questions. Here is a report of one close encounter:

"Hello, are you a practising electrician?"

I give the standard reply. "Yes, and one day I hope to get it right." No response. Not a hint of a smile; must have heard it before.

"I've got a fault on my car."

I give an invisible inward cringe. "Oh yes?"

"Yes. When I press the brake pedal ..." (It's going to be a good one!) "... the left-hand indicator comes on and the sidelight turns off."

"Nasty." I've turned off, too.

"Can you tell me what's wrong?"

Well, could you tell him? It is possible to waffle on about faulty earths and the like, but could you tell him how to fix it?

Even if you managed to mentally picture the standard lighting circuit there could be numerous combinations which produce the effect described. So let us imagine there are only two possible faults which create this effect. How can you know which is relevant in this case? Well, you can't; you would not be able to decide unless you could physically test your theories on the vehicle itself.

Should I waffle? No, the problem is so … so vague. I think I will own up straight away and put us both out of our misery. Theory devoid of practice is waffle.

"It can't be done."

The poor man is taken aback by this admission – honest people get nowhere in this world. Fancy an electrician admitting that he cannot solve a simple car fault. It's disgraceful.

"You've no idea then?" he grumbles.

"Too many," I reply. "That's the trouble. It can't be narrowed down or pinpointed without seeing the car."

Having too much choice is a problem. What is generally needed is a method to narrow down our choices, so we do not waste time on irrelevant issues.

Our works' bus returns at four o'clock and we are slowly transported towards the pale setting sunlight. It is early autumn and the homeward journey is usually a lot warmer than the morning ride. At this time of year there is a complete contrast between cold mornings and warm afternoons.

We soon reach the first town and a couple of fellow travellers alight. We are then treated to another puzzling facet of bus travel just as the bus is about to pull away. An elderly gent taps on the bus door with a walking stick. The driver presses a button, something pneumatic hisses and the doors open. The senior citizen asks where the bus is going, but the driver ignores this and replies: "I'm sorry, sir, I can't let you on. This is a private bus." Funny, it just looks like an ordinary bus to me: dirty windows, grubby floor and half-empty. One would imagine that a bus company known to be losing a lot of money each year would be glad to have people banging on their doors, but it appears not. They are not the least bit anxious to pick up fare-paying passengers.

"Come on, son," the man pleads. "I've been stood here for twenty minutes and m' leg's bad."

"Sorry sir. It's more than my job's worth. This is a private bus."

I have seen this performance a few times. It is another one of those irritating little observations which makes you feel that something unseen is wrong. What is behind it all?

I confess I do not feel any remorse for cheeky kids not allowed on our private bus. It's the old-age pensioners I have sympathy with; elderly ladies, whose legs cannot stand for great lengths of time, are curtly told "private bus" as the doors hiss shut in their faces. The gent with the walking stick probably endured two world wars to make this country great and now he can't even get on a half-empty bus. Bloody great!

What is the underlying cause of this official impersonal indifference we now show to one another? As we progress, things may become a little clearer.

Tracy and Linda are waiting by the bus stop in our small family car and we are soon up and over the hill and back to base camp – home.

Tracy has had an exhausting day. Linda is teething and will not entertain herself for long. She is niggly and restless, and is missing her midday snooze. Her nagging gums and lack of sleep make her even more contrary in the afternoon, by which time Tracy is starting to fray around the edges.

Never mind, Dad is home now and ready to fulfil his promise and "have a good play".

Linda is now at an age where she can move about quite freely by crawling. She drags herself along on her elbows, while her little legs follow behind in an independent manner, similar to a butterfly swimmer on dry land.

To start this evening's entertainment, I build a tall tower, at least two feet high. I balance a ball on a hoop, then another hoop, a ball, another hoop, a fluffy woollen ball, a plant pot and a yoghurt tub on top. An impartial observer would plainly see that I am enjoying this game while Linda isn't too bothered. Sometimes she scrambles over the hairy, woolly rug, like a sniper crossing a grassy field and then she will poke down the tower using her finger as a bayonet. From across the room the unusual combination of toys looks like one larger new toy, but when it collapses into familiar component parts, then the novelty wears off and Linda moves away.

Occasionally, she ignores my tower-building programme and would rather try pulling the safety cap out of our one low-lying electrical socket. Still, I am enjoying it! And I suspect that a lot of things we adults do, supposedly to entertain our children, is a clandestine method of amusing ourselves. In some ways this is a shame, as a lot of adult habits are not necessarily good examples to follow.

You may recall my toy tower (it was at least two feet tall – honest!) contained two items not usually listed as toys: a plastic plant pot and an empty yoghurt tub. Linda has other gaily-coloured toys; some were quite expensive and others are quite elaborate. Yet, the plant pot and yoghurt tub are among her favourite playthings. Her measuring stick is whether an object causes pleasant or unpleasant feelings inside. Sucking an empty yoghurt tub is just as pleasant as sucking an expensive, elaborate toy. To her they both have the same pleasure value, whereas the adults' idea of value inevitably has a financial connotation tagged on to it.

For instance, this summer Tracy and I took Linda to the zoo to see all manner of creatures: sea lions, birds of prey, elephants and bears. For most of the time Linda's attention was focused on an empty plastic cup, previously filled with orange juice to quench my thirst. There we were trying to show her something interesting and she's sucking a plastic cup! Perhaps we had forgotten that when the whole world is new, a plastic cup really is interesting in its own right.

A baby has no preconceived ideas of good or bad, right or wrong – no morality. The adults of this world preach these things, but the world is full of very diverse ideas and the adults sometimes get confused and confusion leads to doubt and then to instability. A perfect synopsis of the resulting mischief is the six o'clock news.

I'm sitting on the floor in a Buddha position when Linda scampers over to play with my slippers. The television is on and the news begins. First item: the latest high unemployment figures, bleak and depressing. Next: inner city street riots. An article about the nuclear build-up in America follows, then a film of South African troops invading another country. Russia stages massive war games near the iron curtain and England's football team lose 2-1 against a team of part-timers from beyond the Arctic Circle; the British fans riot.

This perpetual diet of conflict is really uninspiring stuff. It's enough to make you slide your dumplings to one side of the plate – that is, if you happen to be eating dumplings at the time. It breeds despair and shrivels hope.

Linda pulls my trouser leg in an attempt to grasp the tea tray perched on my knees.

"No, Linda. Play with your toys." Little chance of that, but she's not having the dumplings I've left on the edge of my plate.

I lower her gently to a crawling position and she promptly rolls around and grabs my leg again. I suppose we are playing a new game.

"Linda. No!"

She scrutinises my face to see if I really mean it. I frown; the game is over. Now she grabs the newspaper by my side…

"*Linda ...*" I retrieve most of it, except for the sports page. The England football match and riot report is, quite suitably, torn to shreds.

Newspapers are another source of mass gloom. Bold headlines fuel the fires of conflict, highlighting misunderstandings and accentuating the trivial. Normality is not newsworthy; regrettable but true. Hence the menu tends towards the outer fringes of social activity, far removed from the mundane, everyday existence of the average reader. An exaggerated image of the world is conjured up: a world brimming with perversion, cruelty, hypocrisy and distrust. Of course the world does contain such elements and when presented with conviction, susceptible minds can be easily influenced. People may respond by becoming hardened and callous, their moods matching the projected images of their surroundings.

Most of the newspapers have a political bias which is maintained regardless of how badly their side is performing. The editor dogmatically preaches the faith, delivering a sermon a day like a latter-day Moses. His congregation are drifting through the desert of life, looking aimlessly for the best path to follow for a while and the editor aims to provide some unsolicited guidance to appease his political masters.

It is Linda's bedtime. Tracy takes her to the bedroom to get changed. I watch the weather forecast then wash the teatime dishes, while eight ounces of milk is boiled and cooled for Linda's supper. Later, Tracy starts knitting a cardigan and we chat about the day's events over a refreshing cup of hot coffee. It is time to unwind; to recharge the biological batteries for tomorrow.

It was quite a normal day. I walked to the bus stop admiring beautiful gardens whilst stepping over dog crap and rubbish on the adjacent path. I waited at the bus shelter, contemplating the rubbish strewn all around whilst wondering if the bus would arrive.

In work we glimpsed the frailty of a humble electrician, unable to solve a simple car problem. I travelled home on a half-empty bus that was unable to pick up passengers and finally the media treated me to a scenario of world chaos, reinforced by the pessimistic parlance of the printed page.

That's normal?

Man's science has become the bedrock of modern civilisation. With it he can cure diseases, explore other worlds and harness the immense latent energy locked up within atoms. He can fashion intricate and complex processes out of the Earth's raw materials. And he seems to do this with (if not ease) a certain degree of confidence and flair. His technical ability is no longer in doubt.

Man also desires order. He devises systems to achieve this purpose, but his skill in this area is weak. Unlike a scientific experiment, society is a highly dynamic and volatile mixture with lots of changing variables all over the place; very hard to analyse, even harder to predict. Yet below the apparent surface chaos lies a furtive hierarchy of order, an invisible abstract framework to which our lives are impersonally pinned. Impersonal: that word has immense meanings in the context of the events witnessed during our typical day. All were, in one way or another, impersonal.

I find it a strange feeling to be reading my own composition some thirty years later. I begin to remember where this essay is heading and why, which is hardly surprising, yet there is something about it that seems otherworldly.

Slowly I am developing a theme by uncovering anomalies that are a feature of the problem I later try to explain and correct. Enigmas that prompted me look for a better way in the first place, simple things most readers will identify with.

Meanwhile, I think it worth mentioning that (at the start of my career) we made and measured everything in imperial units: yards, feet and inches. We made apprentice test pieces by hand to tolerances of less than five thou' (five thousandths of an inch), then (somewhere in mid-career) there was a shift to metric units. This pulled us into line with Europe but caused a disparity with the United States of America, which still uses imperial measurements.

I just accepted the change in my stride and later on (when working on lasers) was quite happy to discuss wavelengths of light in nanometres, or even angstroms.

So in this work I might drift between imperial and metric as I do when measuring jobs around the house. As long as the object I'm making fits the cavity, then it's job done.

And so we continue the journey to try to uncover a common denominator to the puzzles we have encountered so far.

Branch 3

Early Learning

Each Saturday morning, Tracy and I take Linda on our weekly shopping spree to the local town centre. Linda is awake at seven-thirty, which gives us plenty of time to organise ourselves before we leave at around ten o'clock. During the tourist season the traffic is usually heavy by mid-morning, so we aim to reach our parking place before the traffic jam commences.

The car park is half full on arrival. After a few moments' thought, I select a suitable spot and reverse into the space. Tracy wanders off to buy the obligatory ticket from a machine. Meanwhile, I unfold the baby-buggy. This is an incredible piece of engineering. When folded it occupies the space of a small golf bag; but undo one clip and it unwinds like a praying mantis in slow motion. Levers and cross members flex and corduroy is stretched; then with a positive "click" a chair on wheels is revealed.

Linda is transferred from the car's safety seat to the buggy seat and fastened in securely. Tracy pushes, I carry the bag.

The centre of town is a hive of activity. A team of council workers is blocking off each end of the main street with temporary ranch-style fencing to create a pedestrian zone free of traffic and full of market traders. Stalls, made of scaffolding and canvas, ring the newly-formed arena and a vast variety of merchandise is on display. Some items spill over onto the kerb, making buggy driving a skilful exercise. Two policemen give the scene an air of authority, occasionally pointing confused tourists in the right direction.

I examine my shopping list. Only a few items today, which is fortunate, for I only have a few pounds to spare. The items are: wooden bricks, a grow bag, and "wood for frame" – a cryptic reminder which I alone can interpret. The wooden bricks are for Linda and me to play with. She is very fond of knocking down the toy towers I build, so now I am going to try her on some more complex structures. I don't know any child psychologists to speak to and I am unaware of the current frame of mind of apparently schizophrenic child doctors, so I am going to try a few ideas of my own and see where they lead.

Tracy and Linda set off to buy meat and a birthday card and I venture forth to the toy shop at the far side of town. There I find a good selection of quality wooden toys. Ah! Those were the days, before Hong Kong discovered plastic. Now the plastic toys predominate, while wood has become a rarer commodity. Hence the prices! Twenty wooden bricks for three pounds seventy-five pence! That's nearly twenty pence per block. Perhaps I am out of touch with the FT Wood Index. A quick mental calculation reveals that buying the bricks will leave me with a pocketful of loose change, no grow bag and no "wood for frame". I leave, not in despair; I am thinking of a way around this impasse.

I see Tracy and Linda waiting at the prearranged place, but as I approach two uniformed men step forward, infringing on to my line of travel. Apparently, they have some connection with the RAF. Why they are collecting I never discover. Perhaps, like me, the RAF exists at a subsistence level.

"Did you see that, Tracy?" I ask.

"What?"

"The RAF collecting for a couple of jump-jets."

"Don't be daft." I am reprimanded. "It will be for some charity. Did you get the bricks?"

I tell Tracy the price and add, "Have you noticed; you never get anyone collecting on behalf of poor electricians?"

We make our way slowly back towards the car. Tracy tells how Linda grabbed a handful of toys in the newsagent's shop. Linda will soon have to learn the meaning of the word "no", hopefully in a way that won't destroy her curiosity.

The girls continue their journey to the car; the shopping basket is unceremoniously hooked over the buggy handles. I detour across the road to the local hardware shop to inspect the wood sections. The wood for my frame is easily found. I check its length with a tape measure borrowed from another shelf. Two racks away I notice a square sectioned wooden post, six feet in length and priced at just over one pound; enough material for at least thirty wooden bricks when sawn. Excellent!

I pay for the wood and a grow bag, the latter on display on the pavement outside. I even get a little change. So, with a mild feeling akin to contentment, we travel home after a successful morning shopping.

In the afternoon the sun shines brightly and I am committed to painting the eaves of the bungalow. This kind of job has to be tackled when the weather is fine, before the onset of winter. I don't enjoy painting. I can tolerate it if it is an act of preservation, but painting for aesthetic reasons is not high on my list of priorities. I am glad when my painting quota for the day is completed and, at last, I can start to fashion the wooden blocks.

Marking and cutting the wooden blocks takes longer than I anticipated, half an hour for ten blocks, and even then their edges are still rough and unsuitable for young fingers. I leave them on the garage bench. Tea is ready. I decide to call it a day.

Next morning the Sun is beating down again, though there is a cold wind. I am eager to finish the blocks, but the painting takes precedence; Linda's bedroom window gets the treatment today. It is mid-afternoon before I can redirect my attention to the rough wooden blocks.

Each block has the corners planed and the sides smoothed with sandpaper until it is splinter-free. This process takes quite a time and it dawns upon me that if the chap who makes twenty for three pounds seventy-five pence has not become mechanised, then he is not going to make much profit.

As I complete each block, I take it inside to show Linda. She is sitting on the lounge carpet. The first block goes straight to her mouth. *"Funny taste, this,"* she might be thinking – not your boring tasteless plastic.

Her reaction interests me, so I sniff and lick the second block myself. Hmmm–yes! A cheeky little wood; a vintage resin perhaps?

The third block is ready. Inside, Linda has moved over to the kitchen doorway and is now sitting behind the safety gate which literally bars her progress into the kitchen area. Her smile stretches from ear to ear as I hold out the third block between finger and thumb and move in her direction. She giggles and waves her arms excitedly, which loosens her grip on block number one. Through the air it flies and clatters across the kitchen floor. She looks at me with questioning eyes; a pause for approval? I smile and she chuckles and squeaks with delight at her new achievement.

When all ten blocks are completed, I tidy the garage, then myself.

Leaving the cold wind to rustle the trees outside, I settle down on the lounge floor in front of a warm fire, ready to build some wooden structures for Linda.

Now the experiment can begin in earnest. First I build a tall tower ten blocks high. Linda crawls over to examine this new shape, but before she reaches it, the vibration of her approach is transmitted through the carpet and underlay and the unstable structure tilts and tumbles. Linda manoeuvres from a crawling position to a sitting stance and grabs the nearest block. I gather the remaining blocks and crawl out of her immediate range to build another shape. This time I strengthen the base by arranging four blocks in a flat square, before building another tower. With a flourish of ostentation, I place the empty yoghurt tub on top.

"Come on, Linda!" I call. "Knock the tower over." It reminds me of a lighthouse, but Linda has never seen one.

She looks around on hearing her name. "Knock", "tower" and "over" mean nothing to her yet. All she wants to do is touch the new toy and when she does Tracy and I cry "hooray!" and laugh.

I return to my original spot, taking nine bricks with me. Linda sucks the tenth block, which consequently changes from white to a darker damp cream colour. I wonder why?

We are going back and forth across the carpet like a couple of crown green bowlers chasing a jack. Linda follows before I start building, so I have to rush to construct an eight-block cube. Linda arrives and razes it to the carpet.

There is only one more shape to try for now. I scuttle back over the round rug in the centre of the room and place four blocks in a row. Next, three blocks on top of them, then two, then one. This forms a pyramid shape – one with a short life expectancy. Here comes Linda! Oh-oh … there it goes.

Now for the final test before Linda becomes fed up with this game. This time there is a square of four bricks and a pyramid using six bricks. In-between I place a ball. These are intended to form some basic geometric shapes, as below…

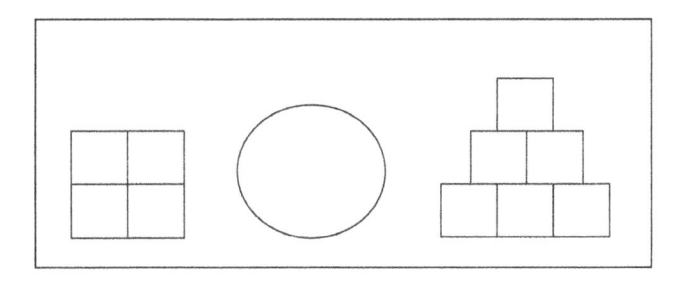

Will Linda show a preference for a particular shape? She soon provides the answer. The square is shattered, the ball bounces out of her path and the top brick of the pyramid is brushed off. Her trailing left leg pushes aside the pyramid base as she wanders away to resume her study of the five pence plant pot.

The novelty has worn off and the new toy, the blocks, become just one more item to fill her toy basket.

Now the experiment has ended I shall explain my thoughts by returning to the early morning walk to the bus stop. As you may recall, the journey was divided geographically; the first half through open pasture land, the second half a tarmac corridor enclosed by man-made dwellings. The first half was largely fashioned by nature; the second half designed by modern man and there is a noticeable difference between the two styles. Nature, in the raw, on the macroscopic level, contains no rigid geometry; whereas man-made artefacts are usually based on a solid geometric shape.

This visual disparity is lessened by two things: first, our familiarity with this type of contrasting surroundings and, second, the geometric taming of nature. In general, fields have been squared and hedgerows trimmed horizontal by mechanical cutters. Vertical fencing posts are neatly spaced and electricity pylons suspend energy on towering stress-bearing angular frameworks. To really get back to nature you need to find an island of ancient woodland and have time to stroll through it. There you will not find anything organic with a solid geometric shape. You might if you looked at the microscopic or molecular level, but with normal vision (through the eyes of a child) you will not see squares, or triangles, or cubes or pyramids. Woodland contains a spontaneity geometry forbids; the trees flex and bend, the grass sways in the wind and tree tops form a curving canopy against the skyline.

Why this conflict of styles?

When we examine nature we are studying a living entity. Meanwhile, geometry's modern definition is "the mathematical study of solid objects".

Man's fascination for geometric regularity is not a hereditary endowment. There is nothing in Linda's genes telling her to build towers, cubes or pyramids, and at present she has no preference for any particular geometric shape. To her a plant pot is as good as a pyramid.

The geometric concepts are put into us after birth. We see them surrounding us as soon as our eyes can focus, then we play with toys based upon their shapes. Eventually, in school, we are taught how to utilise geometry as a powerful mathematical tool. With it we design things; simple things at first, but later, at college or university, complex structures can be created which obey physical laws and behave according to fixed rules. And all this is good.

All this order gives us a feeling of stability. Your house has rigidity which its geometry enhances. Who would wish to live in a building that flexes and changes shape as a tree does?

When (in work) a machine breaks down, I am aided in fault-finding by the knowledge that empirical rules govern the machine's physical construction. Therefore, if certain events happen predictable effects will normally ensue. If the rules for machines were variable I would not know where to start looking.

Man's science and its application – technology – are masters of the physical world. We possess the knowledge to reach out for an earthly utopia or commit suicidal oblivion. But why should the latter option be even worthy of consideration?

Why the disparity? Why the disharmony?

Sometimes the solution to complex problems is so elusive that no answer readily comes to mind. Then one day, when you are not looking for an answer or anything else in particular, an apparently unrelated event reveals an intriguing little puzzle. But life is full of enigmas and we are accustomed to ignoring anomalies so we can persistently be wrong in a more relaxed manner. I had better mention now that I will not be preaching a return-to-nature philosophy in these pages. I just want to explain a novel viewpoint I have uncovered. My moment of insight started by revealing a tantalising paradox: the geometry that enhances the stability of structures, and whose shapes surround us from birth to death, also casts an impersonal and inanimate framework around each of us in a way which is alien to man's intrinsic evolutionary heritage. Rigid geometry and organic life forms do not combine, they collide.

Recently I discovered nine of the original ten wooden blocks Linda and I used to play with. I seem to recall I turned the tenth into Linda's first die, with dots on the six faces from one to six. Finding them brings a lump to my throat. I buy a special box to keep them in for posterity.

Meanwhile, many Moons ago, the town centre was bypassed and Saturday morning shopping became easier to reach. We still visit relatively early in the morning, as the outdoor market makes an attractive venue for the locals who stream in from surrounding villages to see the bargains on offer from the stallholders.

Of late, I've taken to going down into town on the bus and walking back up the hill, about two miles, to keep my fitness levels high. This saves on petrol which is far more expensive than it used to be way back then. Consequently I insist on having a pre-walk treat, coffee and large slice of brown toast from the established café in the town centre.

The next section is more thesis than narrative, though it is a necessary branch we have to encounter along the way. Here I began to think about geometry in relation to the world of work as it was then. Yet there may be similarities to recognise in a variety of present day occupations.

Branch 4

Pyramids at Work

The alarm clock clanks. I am still tired, so I turn over and reawaken fifteen minutes later. Fortunately my schedule allows for ten minutes of unplanned delay. Nevertheless, a few cuts are necessary. I omit to shave [*slob*] and forgo the tea-with-sugar.

Coat on, case in hand and I'm ready.

I pull the front door closed, then realise it is raining. I don't have a hat nor umbrella. Perhaps it will only be a light shower.

Halfway up the hill it pours down; no shelter here. The electricity cables are strung between pylons like huge guitar strings. They buzz and hiss an unearthly tune; a wailing lament which surrounds me and permeates my head. I walk faster.

At the halfway stage I check my watch. Seven o'clock. Shall I run? No, there is no need to panic, but remember the bus did come on time once. I keep walking – fast.

I arrive at the bus stop, damp with sweat inside my coat and getting wetter all the time from the persistent rain. It is ten minutes past seven, the bus arrives at twenty past and we arrive late. Not a good start.

My first job is to repair an air-conditioning unit. This functions on the same principles as a domestic refrigerator. An evaporator collects heat, a compressor concentrates it and a condenser disperses the waste, or unwanted heat. The main difference between the fridge and an air-conditioning unit is the addition, in the latter case, of two fans. One circulates air around the evaporator in the room being cooled and the other forces the unwanted heat into the atmosphere.

The unit has broken down and I am given a work card with a lot of identification numbers on it and a simple request, hand written: "Please repair air-conditioning unit." No clues as to where the fault lies.

The procedure for sanctioning a job is straightforward. The person responsible for the faulty item writes a work card and gives it to my foreman along with a brief verbal explanation of the fault. The foreman gives the work card to my chargehand and tells him his interpretation of the fault and his priority for fixing it. The chargehand delegates the work and advises how best to tackle the problem. And so we can derive from this process a simple structure – the Chain of Command.

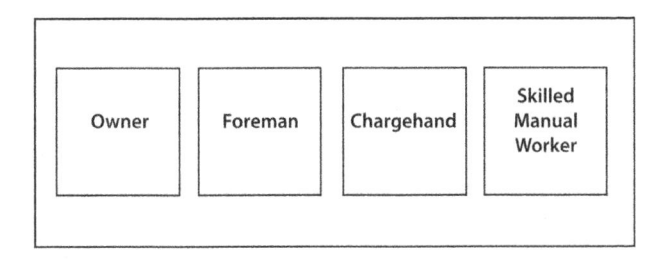

Owner	Foreman	Chargehand	Skilled Manual Worker

In theory, I should now be able to go straight to the job and apply immediate corrective action, and on many breakdowns this is indeed the case. But on more complex jobs the Chain of Command has a different shape; it becomes a pyramid hierarchy. Now misunderstandings occur when people with differing priorities put particular emphasis on their own favourite solution, which may not be the best all-round answer.

My approach is to try to talk directly to the person who owns the job (or who wrote the work card) before commencing repair work. This unofficial communication line is shown on the simple structure on the next page though it doesn't feature in the official hierarchy. It is a good public relations exercise, too, which usually pays dividends.

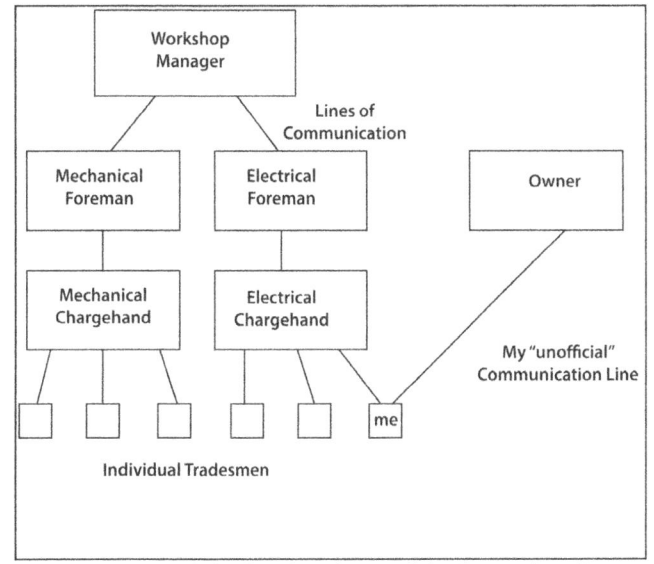

```
                    ┌─────────────┐
                    │  Workshop   │
                    │   Manager   │
                    └─────────────┘
                     /      \  Lines of
                    /        \ Communication
       ┌──────────┐  ┌──────────┐      ┌──────────┐
       │Mechanical│  │Electrical│      │  Owner   │
       │ Foreman  │  │ Foreman  │      │          │
       └──────────┘  └──────────┘      └──────────┘
            │             │                  \
       ┌──────────┐  ┌──────────┐              \
       │Mechanical│  │Electrical│               \
       │Chargehand│  │Chargehand│         My "unofficial"
       └──────────┘  └──────────┘         Communication Line
         / | \         / | \  \
        □ □ □         □ □ □  [me]
            Individual Tradesmen
```

The person responsible for the air-conditioning unit tells me the condenser fan does not appear to be blowing. This is situated outside the main building in a knee-high metal cage with a wooden roof.

I lock the mains isolator in the "OFF" position and place the key on a clip on my belt.

It is still cool outside, but the heavy showers have now dispersed. I button up my overalls to the neck and then remove the wooden roof on top of the condenser enclosure. The fan's impeller is housed in a round plastic guard; this focuses the air through the condenser's radiator grille. I remove the fan guard and find the inside choked with dried oak leaves. The oak tree is very common in this area.

Oh well; might as well have a clean-up first. A clean job is a happy job, or so they say, and perhaps I can then examine the fan motor without contracting swamp fever.

By the time I have returned to the stores and managed to borrow a working industrial vacuum cleaner, it is nearly time for the mid-morning break. I collect my new book, visit the washroom and change back into "civvies".

In the tea-bar I select two pasties and a packet of cheesy biscuits, then order a mug of tea-with-sugar. I settle down in my usual chair, eat the pasties, dunk the biscuits, drink the tea and open the book ... *The British Constitution.*

On page one is the publisher's name. Page two has a preface and acknowledgements and an index: next, a table of English sovereigns. Eventually I reach chapter one, the introduction. The first line proclaims: "Man is by nature a social being," as simple as that, without any qualification. I am stopped dead in my tracks after only one sentence. It is a very general, sweeping statement which strikes me as an extremely naive way to begin a book. It also seems incompatible with man's tribal history and present violent ways. Fortunately, today, I am sitting next to a self-proclaimed student of human nature, Joe. Dare I interrupt *his* studies?

"Er ... excuse me, Joe." He glances up from his studies. "Can I borrow your other paper?"

"Soon finished your book," he responds dryly and passes me his second newspaper. "You don't want to look at these," he continues, "they're full of gloom."

"Exactly," I reply. They are full of conflict.

Let us examine the headlines. Ah! Yes. There are three million people out of work. Meanwhile the Queen's twelve corgis were flown home from Balmoral yesterday. These two articles are side by side on the front page. It's either an amazing error of judgement or the work of an impish editor. Page three has pictures of inner-city riots. The middle of the paper is filled by a murder, a robbery and a large jewellery theft. There is a feature article describing how the extreme left is taking over the country by blaming the media (and newspapers in particular) for all that is wrong. The financial page praises the incumbent right wing Tory government for its bold policies which it then condemns because they do not appear to work. There is a brief review of world news, highlighting all the familiar trouble spots. We finish on the sports page which focuses on the England football team's traumatic trek towards the World Cup. They are still in with a mathematical chance of qualifying if they beat the remaining teams in their group seven-nil and four-two, or if the other teams manage to beat the amateurs from the Arctic Circle. The hooligans arrested after their last away match have now been expelled from that country.

So a synopsis of the news reads like this: unemployment, affluence, riot, murder and robbery, left wing and right wing politics and football – a modern re-enactment of our tribal roots. No evidence of companionship here; little evidence of order, or good organisation either in the country or on the football field.

I hand the paper back.

"I see what you mean, Joe," I comment. "Pretty dismal, eh?"

"No," he replies. "Just normal."

Joe used to be the rear gunner on a Lancaster bomber in the Second World War. Survival rates were low. His perspective must be different from mine. But after a short pause he adds: "Well, slightly worse than normal. Makes you wonder where it will all end."

Now, this last off-the-cuff remark is well worth thinking about, even though it is a well-worn cliché. What goal should we be aiming for? Perhaps if we knew what we should be trying to achieve, working towards it would be a little easier, then we could all "do better". This seems so logical one immediately suspects it is not going to be that easy to find. I mean, every politician is (ostensibly) trying to do his best for the country. So why do they adhere to such diverse, indeed diametrically-opposed ideologies? Are there two such routes to this elusive goal? Surely not, for when the two paths collide there is conflict and when they diverge they are likely to pass the goal on either side. If the paths run parallel, like railway tracks, a degree of stability may be achieved and this creates the illusion of harmony and progress. In fact, the lines never meet, even if the goal lies beyond the foreseeable horizon at infinity.

What I think is needed is a means of focusing man's attention on a goal without actually letting him reach it. We must define the goal, yet make it unattainable: a paradox you may think, but if you attained the goal of life today, then why get out of bed tomorrow?

We cannot send mankind, metaphorically speaking, into a car park with only one space left. There would be no sense of achievement in that. We have got to let him drive around for a while trying to find the best space amongst many empty bays. The car park has to be big, so when mankind stops he will not stagnate; soon he will be looking for somewhere better. Having created an image of a goal, we must make it easy to comprehend, yet expansive enough to free curiosity. Then the masses, confused by the present arguments, counter arguments, innuendo, terminology and lies, can understand the goal and teach it to their children.

Pretty big task, eh?

The tea-break ends. I take the industrial vacuum cleaner outside and park it adjacent to the blocked condenser. I plug the cleaner in and switch on. The nozzle is soon blocked with dry brown leaves. I remove them from the orifice by hand and throw them aside, then plunge the nozzle back into the depths of the condenser. It blocks up again. This time as I remove the blockage I notice it has four legs and − aaarghh! − a tail! It's a dead mouse! Fortunately I'm not squeamish. Another dead mouse and a deceased sparrow follow: must have climbed in to keep warm. And no wonder, the motor is burnt out. I will have to fit another one (if we stock a spare) and I will fit some wire mesh on the front to keep the mice out, too.

I am fortunate my job involves a good variety of work. Some of the work is predictable, without verging on repetitiveness, and sometimes there is an unusual fault to tackle, which occupies your mind totally, like an intriguing puzzle. The jobs themselves are not worth documenting, but the approach to the jobs and the methods employed in fault-finding are the key to success or failure.

One of the most interesting jobs is plant commissioning. This involves the functional checking of a rig to see if it operates satisfactorily and to specification.

In rig building we encounter a management decision-making system, a multi-levelled hierarchy of organisation, far more complex than the simple Chain of Command we have seen already. Indeed, that was only a portion of the base of a pyramid structure made with a lot of boxes and many layers. When commissioning a rig you sometimes find yourself having to think your way right to the top in order to reason out an ambiguous feature. I hope to illustrate this using a simplified pyramid with only four layers which are common to all large industries, yet not specific to any particular one.

```
          Customer
         Design Team
         Management
      Skilled Manual Worker
```

At the top of the structure is a knowledgeable team, committee or even an individual, who has access to funds and a commitment to build something.

That seems fairly reasonable; however, the rest of my analysis is, I confess, decidedly partisan and can be considered a worst case scenario.

The design team take the customer's requirements and complicate them. They never underspend and are committed to preserving their status. The only way they can demonstrate their flair is by elaborating the basics. It's an outlet for them; a means of expression. They rejoice in the complex design and fear the simple solution. Only they can understand the finished product because only they have access to the data, which they closely guard like an insurance policy, in case things go wrong. The design team are theoreticians. They live in catalogues. Sometimes an item they order turns out to be bigger than the catalogue implied. It won't fit. This is a setback entailing the tedious business of altering a drawing.

I will leave out the management layer for now and proceed to the pyramid base. Here we encounter a skilled manual worker. Not so long ago he belonged to the socio-economic group which comprised thirty per cent of the working population. Now both he and the working population are in decline. When a component reaches him that won't fit, he does not worry about it. He used to be keen, when he was an apprentice, but now he knows that, in general, the pay is the same for changing lamp bulbs as it is for wiring complex electronic panels. His take-home pay is his main objective. He has access to materials and a commitment to keep his pay in line with inflation. He rarely does any serious thinking; there is no need in a pyramid system. If an item doesn't fit, then word is passed along the Chain of Command to the design team who sometimes think of a solution. They then "update" the drawing.

The skilled manual worker deals with all the practical aspects of the job. His access to information is limited and so is his knowledge of the job's background. This is a pity, since, in the end, he has to make it work and repair it when it breaks down.

In between the theoretical animals and the practical people is the management layer. They act as an interface between theory and reality. They are trained to prefix every sentence with a cautionary qualifier: "hopefully"; "almost"; "fairly certain"; "possibly". They never state a fact. They know there are none. They act like a referee separating two hostile teams, theory versus practice. At one end of the pitch they receive advice from the home crowd, at the other end from the visiting fans. Meanwhile, in the stands and executive boxes, the administration personnel advise them how to apply the rules.

A good manager is all things to all people. His sports jacket hides a chameleon personality which mirrors the verbal opinions of his immediate surroundings. Once he wanted to change things for the better. Ten years ago he argued passionately for the progressive things he believed in. Now he is content to maintain the status quo. His forte is consistency. He maintains his position, even when some decisions seem to others to be logically wrong; narrow-minded in the country sense and short-sighted in the factory sense. Canute-like, he holds up his rule book against the tide of progress.

At the end of the day, these four differing layers must combine in a way which produces a reward. Usually this is (of course) financial profit.

There is another way to consider this structure. Imagine the four layers as four levels of thought, as below:

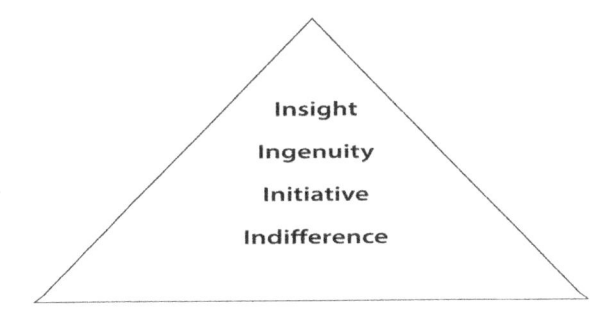

Insight
Ingenuity
Initiative
Indifference

The top level must have insight: a clear understanding of the objectives, coupled with a strong power of discernment.

The second layer (the design team) needs ingenuity: the skill to combine and reason out new ideas.

The management layer needs initiative: the mental ability to organise and process data and the drive to push the project onwards. Finally, with present pyramid structure, the skilled manual worker is obliged to be indifferent. Surprisingly enough, this is as hard to cultivate as insight. Generally speaking, it is not natural for human beings to be indifferent to anything. The skilled manual worker must come to terms with the fact that he sells his skills to an employer and must obediently obey the most illogical directive. Any other course of action generates frustration, which in turn fuels conflict. He must be taught to realise that he is not at the pyramid base holding the rest up. On the contrary, the rest are holding him down in his place; for if the pyramid starts to crumble, he will be the first to go.

Conflict usually starts with a simple misunderstanding. When compounded it leads to a loss of confidence, then a slide towards indifference. Frustration sets in and finally people become sheer bloody-minded.

At each stage along this process there is a loss of respect running in parallel. At the outset there may be some respect, but by the time we get to sheer bloody-mindedness, there is no respect left.

In my experience, people only work well for others whom they respect. They do not work to their capacity in any other circumstances. Financial incentives may help, but there is no need for carrot or stick if you are pulling towards something in which you believe alongside someone you respect.

Pyramid structures do not take any of this into account. Each layer represents a mental level. Each block is occupied by an individual mind and this is where the trouble starts. It is the conflict between one's id, one's individuality, and its rule-bound container. You can try to pigeonhole people but there is always some aspect which sticks out of the box and imposes itself on an adjacent box.

Yet the pyramid remains stable. Its form is difficult to alter. It is inflexible. Inside, the individuals exist in closed compartments which stifle their curiosity and stunt their growth. Rules become everyone's paramount concern, whereas the reason for the rules fades into oblivion. An example: an electrician with fifteen years experience is not allowed to change the mains lead on the oscilloscope he's been using all day because an oscilloscope is an instrument and instruments are serviced by Instrument Mechanics; even though he was just trying to help. Another example: a half empty bus is designated "private" and therefore must not be allowed to pick up normal fare-paying passengers alongside the workers; even if they have fought in two World Wars to make this country great – great. The thinking is so mechanical, so ... so geometric, it's frightening. We are slowly conditioned to fit into the pyramid from birth. The teaching is subtle; subjects are taught in discrete blocks: a history block, a mathematical block, an English block, learning by rote. Girls: a cooking block; boys: a woodwork block. And while we think we are learning subjects, we are really being standardised, as if cloned, so that any cavity appearing in the system can be neatly filled. We are never taught all blocks are interrelated. We are never told some blocks have blurred edges, crumpled edges or edges that overlap. Blurred edges means untidy thinking – naughty boy. This is the rule. Oh yes, there are one or two exceptions, but try to ignore these. Try to write the sentence a different way. Tackle the problem from a different angle. Change your approach. Never admit the method is wrong. Keep on discriminating, isolating, categorising, specialising. There is no room for blurred edges in logical thought. When a fact doesn't fit, state why emphatically, then ignore it until it goes away. And do so consistently to prove you are right.

What is needed is a system of thought to harness our individuality, set it free from its container and focus it on a goal. To instil in people a feeling of purpose, give them a sense of direction, show them how to evolve and how to interrelate with their surroundings.

Can we improve the existing system, or should we aim to destroy it? No, not the latter, nothing so naive. All that is needed is a gentle revelation in our minds.

Of course, since these times the designations "Foreman" and "Chargehand" have become a thing of the past. Nowadays the two roles have been largely subsumed into one ... the Team Leader. The skilled manual worker is still there but now called a technician. There has been a thinning of the layers in the structure.

In a factory with several departmental teams, the prospects for promotion are diminished. Whereas one could aspire to become a chargehand *or* a foreman, now there are fewer Team Leader posts to aim for.

My first factory employed 13,500 people: the second 3,500 people. Large factories, then common, have become a thing of the past. I obtained job offers from two factories when I left school. I picked the bigger. Within ten years employee numbers were reduced dramatically as big factories were replaced by industrial estates which then evolved to become enterprise zones. Thus the big factories were replaced by dozens of smaller manufacturing units.

Newspapers have remained pretty much the same in terms of doom-laden content. The medium is changing, though, from paper to electronic delivery.

The season turns to autumn as my story continues...

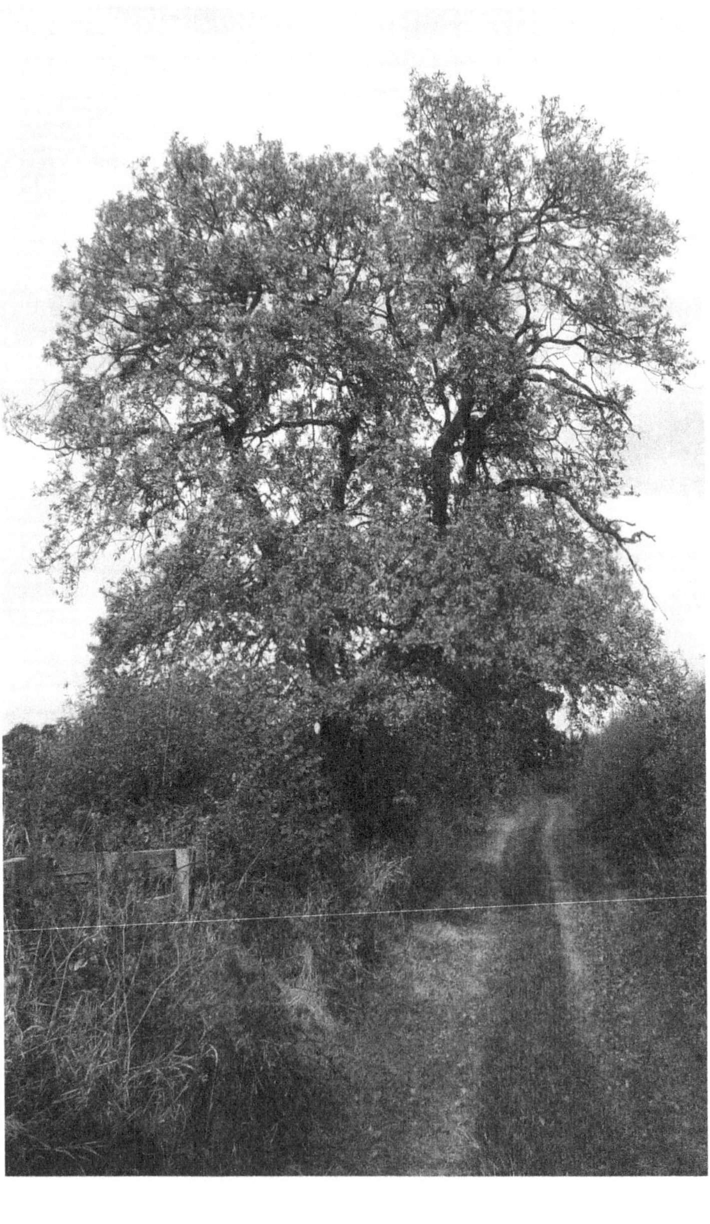

Branch 5

The Birth of Curiosity

One Sunday morning after breakfast, Linda and I visit my parents and my Grandmother. My "Nan" has lived with my Mum and Dad for some years now since Grandpa died. With Linda's recent arrival, everyone is renamed: my Mum and Dad become Nana and Grandad and my Nan becomes Grandma. If you are confused, don't worry, so is everyone else.

Linda sleeps during the car journey, but as soon as the engine stops purring she opens her eyes and looks around. I undo her safety harness and lift her out of the car. Dad (I mean Grandad) appears and carries Linda inside while I gather her bag of toys, fresh nappies and her non-spill drinking cup.

My Mum greets me at the door and we step inside the porch which acts as an airlock; separating the chilly October air from the warm, gas-fired central heating within.

Grandma is waiting in the warmth of the lounge and we are all smiles at seeing each other again.

My parents and Nan returned from a coach tour of Scotland yesterday. I take off my coat and Linda's jacket and we settle down to listen to tales of their holiday.

Now, sitting next to me is a three-foot tall replica of Rupert the Bear, complete with yellow checked trousers and scarf. Linda, who is sitting on the floor at my feet, has noticed this too and is not quite sure what to do about it.

"Look, Linda," my Mum says. She picks up Rupert the Bear and sits him down beside Linda. "It's a present from Scotland."

Linda responds by sucking her dummy – hard. She is puzzled.

I take the dummy out of her mouth. It kept her contented on the journey, but now she is fully awake.

She is still thinking hard about Rupert, especially now he is doing a little jig above her head with the aid of my Dad's right hand. Then Linda makes a decision. She pivots over her left leg on to her palms, thrusts her right leg over her left and heads straight for the waste-paper basket. She remembers from previous visits that she is not allowed to play with the waste-paper basket and this factor seems to double its attraction. It has taken preference over Rupert Bear anyway.

"No, Linda," my Mother states, drawing Linda away by the armpits. She offers Linda an old mail order catalogue, but I suspect that this, too, will not hold her attention for long. Linda paws through the pages. Some are torn, some are crumpled, some are missing; she has had the catalogue before. That is why her interest is waning rapidly.

Now Linda heads for the couch, hoists herself into an upright stance and reaches for the chrome and wooden biscuit barrel at the rear of the corner unit. This is another item of interest which she is not allowed to play with; it's full of toffees.

"No, Linda." It is my turn to stop her this time. I push the barrel out of her reach. Linda stretches upwards on her toes until her little index finger can just tickle the barrel. She cannot grasp it, so all is safe.

Sometimes it is hard to decide whether to let her examine an object or not, since nearly every common object could be misused. On the other hand, she has got to learn about the world sooner or later, so I adopt a flexible approach and judge every situation as it arises.

Curiosity seems to be ethically neutral. It has led to the development of things both good and bad. Where did it arise? How did it develop? Observations of Linda indicate this desire to explore is an inherent human trait. No one has taught her to investigate her surroundings. She is imbued with curiosity; an invisible force, strong, expanding her world daily while her poor old Dad looks on, his curiosity satiated, his hopes tarnished by time. Time – let us return to a time before curiosity existed and plot its steady progress. The pathways are so old they are vague and sometimes untraceable. A modern guide, like me, must confess in advance the trail I envisage may not be totally accurate. I have no desire to mislead you, but the true trail is still unknown.

Ten million years ago, ape-kind was merrily swinging through the trees of Africa when the climate started to change. It became warmer and the forests gradually gave way to open grassland – savannah. Thus ape-kind was obliged to spend more time at ground level looking for food, making them more vulnerable to attack from predators. This encouraged them to stand upright and look around more and this, in turn, released their arms to concentrate on picking fruit, berries and leaves. Ape-kind brought two arboreal assets down to the savannah below: an ability to grasp objects and binocular vision to judge distances accurately – very necessary when jumping through the tree canopies. By around five million years ago ape-kind had evolved into ape-man, a creature that could stand upright and dexterously grasp suitable vegetation selected from the surrounding greenery by his keen discriminating vision. To coordinate all this activity a larger brain was required, larger than needed in the treetops. Then it was just swinging from tree to boring tree, eating whatever they found. Now, new food sources *needed* to be found. Ape-man's physical inheritance was well suited to exploring a new environment. The physical means to examine one's surroundings is a prior qualification needed if curiosity is to flourish.

It is said that "curiosity killed the cat", and indeed, some animals (other than man) do investigate their habitats. But the degree of curiosity is limited by the physical restraints of the animal in question. A cat cannot hold an implement and manipulate it purposefully. No cat has ever used a stick as a tool, that is, as an extension of its physical ability. Let us consider a limit case and consider if a worm has curiosity. Here is a creature with very limited physical ability. Indeed, a worm does not need curiosity to survive. It merely slithers around, usually below ground, digesting decaying plant matter. Who is better evolved? The worm, unchanged for many millions of years, or mankind who has been compelled to adapt to new food supplies? The answer is mankind. Years of inquisitive behaviour have placed him at the present pinnacle of evolution. He can mentally correlate solutions to new situations. If your garden becomes flooded, the poor old worm might drown; the human gardener, however, could devise some means of overcoming the problem. Those possessive of an inquisitive nature seem to set the pace of progress, dragging the remainder of humanity along by a flimsy cord of communication.

Meanwhile, back out of the jungle, ape-man is having (I like to imagine) a marvellous time. He strolls around the grasslands digging up roots, poking sticks into termite mounds and tasting, for the first time, meat, the remains of lion or hyena kills. The quality of his food supply improves and this leads to a slight expansion of the population: a period of stability, a balance with nature, follows.

Ape-man now lives in small social groups of about thirty members. Each member is friendly within the group, but is cautious towards other groups and outside interference. This facet is remarkably still prevalent in today's society; a measure perhaps of how little we really have socially evolved. The common factor linking such a social group today could be quite diverse: support for a local football team, or the comradeship of an electrical section.

Curiosity means "search for knowledge", but ape-man was only concerned with the search for food. Thus the social groups with most knowledge of food resources would predominate over unenlightened groups. Those groups with an eagerness to search for a greater variety of foods were more successful. The desire to explore was an attribute of survival and it became reinforced by successive generations.

By this time, another facet of curiosity was developing. Using trial and error tactics, akin to modern experimentation, ape-man found that another way to increase food supplies was to hunt. Now he would catch and kill, leaving the remains to hyenas – a remarkable turnaround. The kill would be cut with stone tools and shared.

One million years ago ape-man evolved into *Homo erectus*, or upright man. He led a hunter-gatherer way of life, hunting the herds roaming the savannah and gathering fruits, roots and berries as he went. Upright man was a nomad with a brain size twice that of his predecessor. He ate more meat and continued making tools from stone. His wanderings took him out of Africa, into Asia and southern Europe.

Six hundred thousand years ago, in the northern regions occupied by man, the climate turned colder. Great ice sheets descended from the north, freezing and locking up so much water that sea levels dropped dramatically. Land bridges emerged from beneath the seas, linking Asia to America across the Bering Straits, and New Guinea, Indonesia and Australia became accessible. Upright man continued his wanderings into these new regions.

While other animals died of cold, man was stimulated to invent fur clothes and to seek shelter in dry caves. Man now had control of fire. Not only did fire warm him, with it he cooked meat, softening the fibres, making it easier to chew and digest.

During this time mankind's population was very stable, increasing by only one tenth of one per cent each century.

The next phase of mankind's evolution (as classified by modern man) was *Homo sapiens* – wise man. He was fully evolved, biologically speaking, forty thousand years ago; that is, he was physically very similar to present day man. Where he differed was in his cultural evolution. He still lived a hard, rugged existence among a simple social group. Only a change to his life style would expand his outlook.

Eight thousand years ago the nomadic hunter turned to agriculture. He grew wheat and tended sheep and goats. Now his life was consciously tied to a specific area of terrain and his travelling became confined to a smaller region. Man settled down and at this point the population explosion began.

The food supply became more reliable and modest surpluses needed to be stored. To fulfil this objective, certain members of the social group would specialise in the making of pottery and basket ware: the first skilled manual workers.

The second factor needed for curiosity to flourish now emerged: affluence. Before agriculture assured his food supply, the hunter-gatherer had only one priority – finding food. Agriculture released him from the rigours of the past, but this created another facet; resources now had to be defended from aggressive neighbours. Thus guards were organised into armed groups to protect the community's products. The remainder of modern history is still littered with the skirmishes between these tribal cultures, even to the present day.

So curiosity needs two things to prosper: the physical means to explore and also the affluence to nurture it. But there were still obstacles to overcome before curiosity evolved into formalised science. We will encounter some of them shortly.

Three thousand years ago, the early Greek philosophers thought of a few good puzzles for Western man to sort out. In this era Aristotle analysed the world by breaking down arguments into groups or classes of words, a technique which proved the embryo of much to come.

The Greeks also applied speculative reason to the world around them, their opinions differing widely from the theistic view of creation which adjacent neighbours maintained. You could say this was a case of logic versus faith, two views which will never be reconciled.

The Golden Age of (classical) Greece ended around 400BC. Saint Augustine later wrote: "Though the learning of Greece still warms the world at this day, yet they need not boast of their wisdom, it being neither so ancient nor so excellent as our divine religion, and the true wisdom." Rationalism gave way to the Church.

Superstition and tradition impede curiosity. They act as a brake on progress, which can be a good thing in small doses. Tradition provides a degree of continuity threaded through successive generations and is tolerable. Superstition, however, suppresses innovation. It is evil, anti-man, in the same category as dogma.

Around 1500AD a movement called the Reformation emerged. The thousand years of the so-called Dark Ages were coming to an end, yet this was still a difficult time for reason. Any advance of knowledge had to be in agreement with the religious view of creation. Hence a major disagreement occurred when Galileo Galilei published his book *The Starry Messenger* in the year 1610AD. This work argued the Sun was the centre of the Universe and not, as the Church preached, the Earth. Galileo was summoned to appear before the Inquisitor at Florence to be questioned about his theory.

In time, the congregation of the Holy Office, presided over by the Pope, decided to humiliate Galileo. He was taken to be shown the instruments of torture used to extract "confessions", then ordered to retract his heretical views. This he did and with this suppression science around the Mediterranean came to a halt. The scientific revolution moved to northern Europe where men were free to express new ideas.

Curiosity is a driving force of progress, but progress can be good and bad. Curiosity also helped to unearth things like myxomatosis, Agent Orange, DDT and the atom bomb.

Somehow we have to decide, as with Linda, what to touch and what to leave alone.

Mankind's inquisitive mind must be channelled for the benefit of man, not to destroy him. To ignore his inquisitive nature is to suppress it and doing that is a tragic waste.

We saw earlier how a factory's pyramid structure corresponds to four levels of thought and how, at the pyramid base, the skilled manual worker had to cultivate indifference or become prone to frustration. But being indifferent is contrary to mankind's evolutionary heritage. Yes, some can achieve it, but the majority cannot. Millions of years of probing, searching; eons of time, testing, questioning, reasoning, are trapped in our make-up. Curiosity pervades our instincts.

The pyramid structure ignores this heritage and compels us to perform tasks in an inorganic, machine-like fashion. It looks neater on paper that way: specific rules; bar graphs; flow charts. Boundaries are drawn splitting one man's work from the next, when both might benefit from occasionally crossing the fence.

Yet there is an even bigger dilemma inherent in rigid pyramid hierarchies. The pyramid breeds indifference at the base, then, with the passage of time, vacancies occur on a higher layer. Promotion happens and the successful candidate moves up to the next level, taking his qualifications with him – the latter including his training in indifference. Ultimately, particularly in large industries, this can invert the thought layers in the pyramid. Now near the top we may find indifference, while the ingenuity lies dormant at the bottom. Insight has usually disappeared long before this extreme worst-case situation is reached. The lower layer, now containing the ingenuity, is not able to flourish under these circumstances because it is starved of resources. No affluence, weak curiosity, means no progress. The people working on the process or equipment may see better methods to achieve the goal, but are now in a position where they cannot test their ideas or openly discuss them. Their superiors sometimes frown on any idea they perhaps should have considered themselves.

The only true test of an idea is to try it. This requires ideas, but also a management prepared to accept a possible failure to achieve a worthwhile potential gain. But in a diseased pyramid, with indifference near the top, the art is to talk new ideas to death, until they or their lowly inventor go away. It is suppression of a subtle kind.

Meanwhile, elsewhere, pure research still struggles on. Perhaps one day man will catalogue and investigate every aspect of this fascinating little planet and store all the facts on a giant computer, ready for instant recall.

When we then know practically everything there is to know, will we have reached our goal?

Now we have the internet and World Wide Web which largely does catalogue every aspect of the planet, though I don't think this has helped us a great deal in search of an overall goal. It is all still up for grabs. A direction is needed. More analysis is necessary.

Branch 6

Job Analysis

Six-thirty in the morning is definitely not the time Mother Nature intended me to rise. I can feel it is too early. This timing is a by-product of the Industrial Revolution; we all have to reach the factory together to perform our allotted tasks in unison.

I move a curtain aside and glance out of the window. Even the Sun isn't up yet.

I follow the standard routine as on many other days: wash, shave, eat, leave. The factory has mechanised my behaviour before I reach its gates.

Outside the air is cold and damp. I pause to fasten the top toggle on my duffel coat. Soon I leave the estate behind and head for the open road. A light drizzle starts to fall. I stop once more and hoist my duffel coat hood up and over my head.

The dawn is signalled by a crimson glow over the eastern horizon, while dark purple clouds loom from the West. Fresh rain, distilled from the oceans, cleanses the earth and dampens my trouser legs. An awesome battle is about to re-enacted. A fight between opposing elements: fire and water; a meteorological conflict.

I imagine a heavenly ringmaster setting the scene: "My Lords, Ladies and Gentlemen. Today's contention is between two old adversaries: on my left, in the red corner, that fiery inferno – the Sun!" [Cheers]; "on my right, in the gloomy purple corner, the bringer of dark weather and damp trousers." [Boooo …] "Preventer of clothes-line drying; creator of mottled vision through one's glasses – the rain!"

HSSSTTtttzzz … sing the high voltage cables dangling above on pylons, as if echoing my disapproval of the steady rain.

I reach the top of the hill and look down on a valley filled with the orange glow of sodium-vapour street lighting. As I walk down the path, the artificial light is reflected by the telephone trench top, which shimmers like a river of gold. The raindrops coagulate into rivulets washing into the gutter, forming a liquid honey stream that cascades down gaping grids; down, always downward, until, in some far-off chasm, it reunites with the oceans and is reborn again.

White cars become golden cars; fir trees drip tinsel from their needles. The bungalows and driveways, lawns and hedges all assume an orange hue, all except for a lonely black cat, stubbornly remaining black against the orange backcloth, a testament to his mythical magical power.

Suddenly the scene vanishes, terminated by the emotionless click of a pre-programmed time switch. The street lights go off and I become plunged, once again, into the cold wet reality of a grey autumn day.

I march on.

There is no shelter from the rain. In summer, the occasional tree would provide a brief respite, but now the oak and beech trees are almost bare. Their skeletal forms are silhouetted against a blood red sky, giving them a royal, majestic quality, as if they are imbued with some unknown purpose. Yet the life force within them remains dormant until the Earth's axis resides on a more favourable inclination. Rust-brown leaves mottle the footpath and collect in windless pockets.

When I reach the bus shelter, the weather gives me no choice but to stand amongst the refuse. Soon the working day will officially begin. As soon as I put my foot on the bus I become a part of the Great Factory Machine. The bus is a satellite containing a microcosm of the factory's personnel. The paramount item we have in common is the factory, so thinking and talking about it starts here – on the bus.

For a change, the bus driver takes us on a more scenic route today. He abandons the new dual-carriageway bypass and travels down old meandering roads, past farmyards and cosy looking whitewashed cottages.

The rest of the day is quite conventional. Clock on: change into overalls, fix something electrical and have a break. After break I'm working near the top rung of an old, solid, wooden ladder, when a passer-by (not in my Chain of Command) tells me the inspection label on the ladder is out of date and I am therefore uninsured in the event of an accident. The rules state he is perfectly right. I lay the ladder down, inform my foreman and get another ladder from the stores. This has the correct date stamp on it, even though it wobbles and is flimsy by comparison to the older ladder. Knowing you are insured on top of a wobbly ladder gives only a vague feeling of comfort. I felt twice as safe on the older version.

In the afternoon I fix something electrical, have a cup of tea, fix something electrical, get changed, clock off and catch the bus home. And it was on just such a typically dull day, when my spirits were low, my trouser legs damp and my ladder wobbly, that depression set in. I experienced a nasty bout of the what-a-waste-of-a-lifetime blues ... Oh yeah! I had collided head on with the going nowhere syndrome, a problem a lot of people, if not all, face at some time in their careers. A stage is reached where you feel you have outgrown the job. One feels (at last?) that perhaps you could do better after all.

Tradesmen are practically minded people who deal in material matters. A screw either fits or it doesn't and if it doesn't we get another one which does. We seek yes and no answers, black or white decisions. Very few answers are in this category and sometimes we are upset by the grey responses of a diplomatic management. So, when we feel angry, frustrated, upset or depressed, we blame *The System*. And we persist in doing so until the myth becomes a reality in our minds.

The System is wrong, you cannot change it and that is that: end of story; full stop.

But it wasn't like that for me. Not quite. Because, at least, I did want to understand The System to see if condemning it outright was justified.

Whilst going through this phase I decided not to get angry, frustrated or upset and resolved to remain quiet and introspective while I analysed the phenomenon.

At first I was only concerned with my electrical career, but the diagnosis blossomed into a way of looking at things, a philosophy if you like, which I found appealing. But I am getting ahead of myself here. I want you to see how these ideas developed so you can understand my reasoning later.

I had noticed my short career was similar to those of other electricians in the factory; few of us now worked for the same firm where we had served our apprenticeship. There seemed to be a desire to leave one's place of training to "see the world".

In those early days we thought there must be better factories elsewhere and our careers seemed to follow a common pattern regardless of the individuals involved, as if a career came in phases and as each phase ended a choice for the future was offered. So far I had identified two phases common to most careers. I called these the adolescent phase and the fully-trained phase. Then I drew the following diagram to represent them:

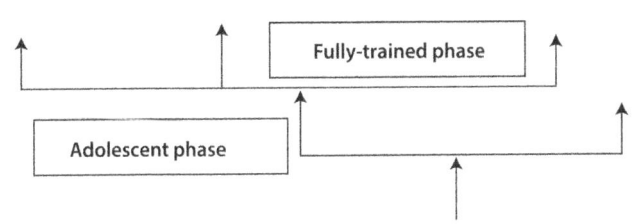

During an apprenticeship we undergo an adolescent phase of learning. The apprentice at this stage thinks (some know) he is going to be the best electrician in the world, but not if he stays in "this old factory".

He has too much talent to merely learn how to fix things. He is wasted. He should be out there (somewhere) designing the perpetual motion generator, or inventing the replacement for the electron. But where out there shall he go? "Anywhere is better than here," is the reply.

While he mulls over the problem he reaches the end of his adolescent phase. A number of options are open to him and he drifts along (apparently laterally) trying to make a decision. His efforts are hampered by three main things. He does not know (a) what is a good career; (b) what is a good electrician and (c) the goal of life. So he chooses, virtually at random, and becomes, let's say, an electrician in another factory.

Stripped of the urge to migrate to greener pastures, he now enters a period of stability where he can apply himself diligently to the task of becoming the best electrician in the world. At the end of this phase he is a fully-trained, competent electrician – hopefully.

His ambitions are now tinged with reality. He would like to be the best electrician in the world, but he knows management are not going to let this happen. When he becomes too competent he becomes a nuisance in their eyes: always experimenting with new devices, the state-of-the-art technology, which they do not understand. He may complain about the lack of modern equipment, which they are out of touch with and won't order; or he moans about the lack of information, which they possess and are hanging on to. Management define a good electrician as an obedient drone, compliant to their every whim. They don't want tradesmen who ask awkward questions, who question decisions. They don't want workers to get emotionally involved with the job. In fact, they don't want workers to become excessively interested in their work, in case they inadvertently cross into management's domain. The workforce can sense this atmosphere immediately.

In this environment the skilled manual worker resides. He has little choice. Potentially he could still be good, creative and skilful, but the longer he resides beneath the pyramid structured management, the more these facets become atrophied. The older he gets and the longer he is subservient to *this* system, the less interest he shows and this reflects in his workmanship.

Meanwhile, our hypothetical electrician has not yet reached this stage. Remember, we were at the end of the fully-trained phase and once again options are available. He has to decide whether to drift along into more uncertainty, or (remembering his previous lateral slide) try to impose some order on the problem.

Let us have a look at the last portion of his career fork:

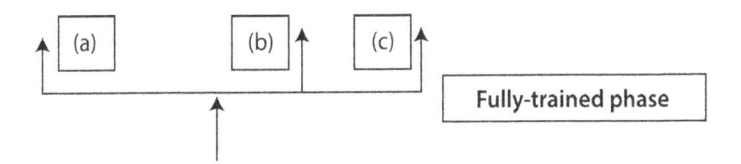

```
   (a)        (b)    (c)
                            Fully-trained phase
```

Now at the end of his fully-trained phase we will assume he has three options open to him. These could be, for example: (a) promotion, (b) further education or (c) move to another factory. Which is the best choice? How can we determine what is a good career?

The trouble with analysis is you always split the question into bits; there is no other way to do it. Ah! But then you concentrate on the bits and forget where they came from. The individual pieces may be solvable, yet it is their relationship to the whole that is of utmost importance. For example, an economist would mainly use money to explain or control human behaviour. Nevertheless, isolating one factor of a materialistic society, even the most important element, does not throw any light upon the question of where that society should be going or what it should be aiming for.

What is needed is not a fragmented study of the jigsaw pieces, or what the jigsaw is made of or how much it cost, but an overall vision to see what the finished jigsaw picture will create.

For these reasons, I felt the questions I asked about a career were invalid. It seemed fundamentally wrong to take a career and analyse it in isolation from the rest of my life. So, how important was a career?

To all but a privileged few, a career is a necessity. It earns them the money to survive. Cynics might think the unemployed survive quite nicely, but this insults the majority who genuinely cannot find work.

To get the problem into perspective, I calculated the amount of time allocated between work and family. In a normal week I spent forty-five hours in work, sixty hours with my family and sixty hours in bed recovering from the first two items. In round figures, work occupied over forty per cent of my active time, a considerable portion. Extrapolating over a forty-year period produces an amazing seventy-two thousand hours in a career. It is a lot of time if wasted. Logic dictates the only reasonable approach is to try to enjoy your career. Spending so much time hating a job is akin to masochism. One path to a happy career is to apply yourself to the task of becoming good at your job, even if, like me, you just drifted into the subject when you left school.

Show me a man who is good at his job and I will show you a happy man. Not always of course; nearly always. If a career was everything in life then I could omit the qualifying sentence that suffixed my statement. But, as we have seen, the other major factor in one's life is the family. A man with family troubles will be unhappy even if he is the best electrician in the world. Thus, a career cannot be truthfully examined as an isolated entity. We must look at the overall picture. But how do the career and the family interrelate?

When I followed my career fork backwards in time, I found it evolved from experiences gained during my formative years at school. My career began when I left school, but the ideas, and the type of career available at the time, were focused by the school environment. In those days local industry needed craftsmen. They visited schools to fill their vacancies by explaining the benefits of their particular training facilities. It was almost as if you were obliging them by signing on. I was accepted by two firms and had to choose between them; what a change from the hopelessness of the present day youngsters.

Anyway, school was an influence on my career.

School is also the place where most people develop a healthy social outlook and could therefore be considered as the starting point of another fork of one's life, leading towards a family. Initially, in school, we chose friends in a stage I will call the gang phase. Later, one may associate with a specific friend more than others. By going to discos, dances, parties, or by playing sport, or in a working atmosphere, a person may be attracted to members of the opposite sex, then perhaps to a special, particular one. If all goes smoothly they get engaged and married, or just live together. They have a child and become a family.

At each phase in this progression, options are open to the individual and suitable selections have to be made.

These two aspects, career and family, have a common denominator – the school years. This influence is at the beginning of both forks. Now the two forks can be drawn together from a common point. For clarity, I have highlighted the main decision paths through the forks in the drawing on the next page.

The simple career fork and the family fork now take on a familiar shape reminiscent of the silhouetted trees I saw earlier in the day. A tree structure – I liked the thought of that. After all, a tree is a living structure that grows just like a career or a family should grow. It seemed to be a structure which favoured analogy with one's own life.

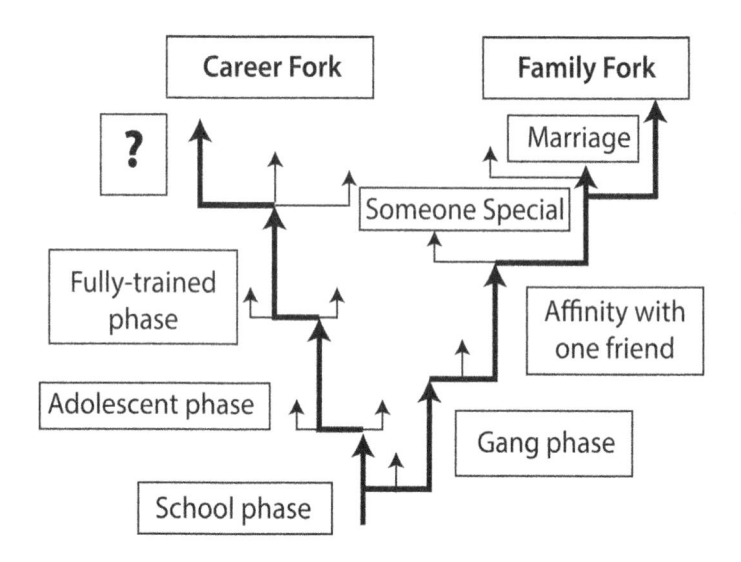

```
                    Career Fork          Family Fork
                        ↑                    ↑
              ?    ↑    ↑              Marriage
                        ↑              ↑      ↑
                        Someone Special
                                ↑
        Fully-trained     ↑           Affinity with
           phase                        one friend
                                ↑
        Adolescent phase  ↑  ↑
                                         Gang phase
                    ↑   ↑
                School phase
```

And there is another attractive feature of a tree structure; unlike pyramid hierarchies, which are mental abstractions, there were many real living trees for me to examine. Perhaps, by inspecting them it would reveal an insight into our origins, our goal and what a good tree looked like. Then we would have something for which to aim.

I made a mental note to study trees more closely. It was a time for re-observation, as if seeing, once again, through the unclouded eyes of a child.

Trees are so familiar near where I live no one gives them a second thought. They're just there: always have been, I suppose. But when you afford them a second glance, when you really take the trouble to look, there is a sense of wonder. Here is a life form which existed before man, before pyramids and will still exist after your death. Could I find in them the revelation I quietly sought?

Although I explained the feelings of hopelessness in a factory environment, perhaps you (dear reader) may have similar feelings in your office or workplace and recognise the same divisions. You may even recognise a similarity with phases of your own life experience.

In the last two chapters we have seen mankind evolve from apes. Likewise, my ideas of tree structures have started to evolve from simple life-path choices to more complex interrelated structures. Evolution seems to be an inherent part of life, though I have never seen Darwin's Theory of Evolution by natural selection expressed mathematically, even though it is a cornerstone of science. Evolution is more of an organising principle.

I believe the Universe knows nothing of mathematics and, like evolution, its fundamental law may not be expressible mathematically. Mathematics, to me, is just a man-made way of describing events, a quite precise way, unless you hit infinity. It's a shorthand way of compacting big, complex ideas full of variables. It may be possible to express what has happened *since* the Big Bang mathematically, meanwhile the fundamental constants of nature (about nineteen of them) have not emerged from any known formula to date and the two bedrocks of science: Einstein's relativity and the quantum, remain irreconcilable. Only five per cent of the stuff in the Universe is actually matter, the rest is presently called dark matter (twenty seven per cent), or dark energy (sixty eight per cent). Of the five per cent we can see or speculate about, more than four per cent is inter-stellar gas. That means all our theories have been derived from less than one per cent of the visible stuff out there and surrounding us. In the meantime we are screwing up the planet; so let's not get too carried away with our omnipotence just yet.

Branch 7

Autumn Advances

In late autumn, when the leaves have finally fallen, I undertake the task of tidying my garden. The front section is relatively tidy since most of the leaves are blown by the prevailing wind onto the back lawn. There are four large trees bordering the drive: three oaks and a sycamore. This may give a false impression of a huge country mansion, whereas the reality is a modest two-bedroom bungalow with a longer than usual drive.

The verdant green leaves usually wither at the beginning of October as the trees withdraw the chlorophyll and sugar back into their trunks ready for winter. The tannins remaining within the leaves colour them yellow, through red, to brown. The trees are usually bare by the end of November. I remember my first autumn in the bungalow. I regularly tidied the drive, religiously brushing the whole drive at least once a week. Now I have learned the best technique is to let nature take its course. Let the wind blow and the leaves fall, then when they are all down and the back lawn has disappeared, it is time to get the leaf rake out of the garage.

The rear garden backs out on to open fields. The neighbouring farmer crops the grassland twice a year and stores it as silage to feed his dairy herd in the winter months. A low-lying hedgerow, interspersed with more large oak trees, leads down to a small copse at the far side of the first field. The oak trees along my drive mark the continuation of this hedgerow as it was in bygone days, before the estate was built. As you can imagine, I have plenty of subject matter for my investigation into the structure of trees. I also have another big advantage on my side – I am not an expert. Therefore, everything I find will need searching for. I hope to uncover a better system in the tree structure than the prevalent pyramid hierarchy, but success is not assured.

There are many things to consider as I rake leaves into a pile on the rear lawn. Hundreds of crispy, crinkly, sun-soaked leaves bury the sparse greenery of the damp grass below: it is like wading through a breakfast cereal for giants.

I use two main methods for disposing of the leaves: a bonfire and a compost heap. The crispy dry leaves on top of the lawn are committed to the flames, while the damp slippery morass below is composted behind the empty, and now unused, oil central-heating tank. The leaves that collect on the vegetable patch and borders are dug into the soil and reintroduced to the earth to continue nature's cycle. And there are plenty of leaves left to send to my father and father-in-law, so they, too, can improve their tree-less soil.

I pause for a rest, lean on the rake and look up at the tall sycamore above. How can my analogy of a tree of life compare with the real thing above me now? There are obvious differences. Unlike my simple drawing of a two-dimensional tree, the real thing is big, its branch structure appears complex and, of course, it is three-dimensional. The latter aspect intrigues me, so I walk up the drive to view the branches from a different angle. As the viewing plane alters the branches appear to intertwine. Two particular branches will relate differently depending upon one's viewpoint. Another difference is the flexibility of the real tree compared to a static drawing. The sycamore sways, branches nod and wobble, the twigs swish about. When the wind blows from one direction certain branches absorb the force. When the wind changes course other branches disperse the aerodynamic energy. The tree's structure is very flexible at the twig end, but semi-rigid down at the trunk; and this flexibility, this motion within limits, I find most appealing.

Tracy calls to me down the drive. "Are you all right? You've been looking at that tree for five minutes now."

"Yes, fine," I respond. "Just counting how many leaves are left." Well, she wouldn't believe the real reason: who would?

Linda is held aloft in Tracy's arms. She looks upward at the trees. The wind generates a hissing noise among their branches. I return to the leafy lawn and the two girls retreat into the warmth of the bungalow.

When the lawn is clear and the compost heap high, I decide to dig the vegetable patch. It is only a small triangular area and it too is pebble-dashed with desiccated leaves. Digging helps the soil replenish itself; it also buries the leaves before they blow back on to the lawn. It seems incredible that a cupful of seemingly lifeless soil can contain millions of protozoa, bacterial and fungal plants; a silent world of activity which breaks down minerals and humus into nourishment for plant life. Digging helps these micro-organisms to breathe and also permits and encourages the easy transition of plant roots through the medium.

My soil is quite clayey, so I use my fork to penetrate deep into the ground. Later in the year I add ash from the solid fuel boiler and sprinkle the ground with lime. This chemically fragments the tilth.

Occasionally the fork gets caught in the subterranean web of roots radiating out from the sycamore tree looming overhead. Most of the roots have been disturbed before. The first time the soil was turned (from virgin pasture) I encountered a tangle of fibrous roots. Now they only ensnare my fork when I dig deep.

Most people know that trees have roots to supply water and mineral nutrients, even though the roots are rarely seen. The last time I saw tree roots on a large scale was in the English Lake District. Down by the shoreline of Lake Windermere the root system of a large tree had been exposed by the fluctuating level of the lake water. I remember posing for a photograph holding one of the grey woody roots, as if I was supporting the tree itself.

Roots can penetrate to the same depth as the tree is tall. If a tree, like an oak, were unearthed with its roots intact, it would reveal an underground system as solid and complex as that above ground.

Earlier I traced my career and family back in time, but where does the individual, now represented by my tree of life, originate? The answer, to use gardening parlance, is in the soil. In my tree analogy, our heritage is represented by the roots. I say "my analogy" yet this concept is not new and should be familiar to most people, for I do not claim originality in parts of this dissertation, only the fundamental authenticity of my synthesis when proposed as an integrated whole.

I picture, in my mind's eye, my roots diverging outwards to my parents: my grandparents and so on, until a large complex root system is formed. The roots go deeper. Eventually the roots converge again, so the image is now like a diamond-shaped matrix focussing on our primitive ancestors who roamed in Africa long ago. But the roots do not stop here. The tap root goes deeper still, down into the unknown, down into the historical void until it touches the singularity where the Universe and Time began. You and I, the atoms in our bodies, are linked to that event by roots so deep, in Time and Space, we sometimes forget this fact. Some impudently start to invent reasons for Creation and unilaterally declare we are the pinnacle of evolution, even imagining we are made in God's image.

It's time for another rest. Tracy calls and I go inside for a cup of hot coffee. The daylight is starting to fade now and so is my energy, but as we approach winter it is wise to tackle the outside jobs when the weather is dry. There will be many worse days than today.

My glasses steam up as the warm inside air condenses moisture on to the cool glass lenses. I wash my hands.

Linda is playing with a jigsaw – or is it? There is only one piece. I made it myself. The one-piece jigsaw has a picture of two teddy bears stuck on to a piece of hardboard. The outer edge is wavy. At the moment we are practising putting it down the right way up and since this can be done wrongly, it must truly be a puzzle.

Linda stems from a union of two root stocks, me with Tracy. We, in turn, are the result of our parents' activities and the sequence continues backwards in time. Linda's family forms her roots, which, if they remain healthy, will provide stability while she grows. I reiterate the concept is not new; I am only unearthing this ground to replenish it.

When the jigsaw lesson is over and the coffee cup empty I don my gardening coat and venture forth for one last session before the light disappears. I take the jigsaw with me and cut it in half, right between the teddies, with my fretsaw.

My greenhouse is empty now except for six winter lettuces which look limp and distinctly out of season. The only regular job is to water the first-year strawberry runners to prepare them for next spring. One other job is to collect some soil to fill the strawberry tub.

Down by the small copse, one field away, lives a family of moles. They provide me with lovely, freshly dug, worm-free mole hills, which is good news for my garden, but not for the worms. Take three parts mole hill, add one part sand and a sprinkling of nitrogenous fertiliser and the result is a good tried and tested growing medium. Not too good for cuttings (I don't know why) but excellent for seedlings and strawberries.

I collect a plastic bucket and a hand trowel, and then undo the loose fence behind the greenhouse. I launch an old scaffolding plank across the muddy ditch and my drawbridge to the field beyond is complete.

The transformation from the confined garden to the large open space is an almost magical experience. It is not quite nature-in-the-raw, more a case of nature medium-rare. The work of geometric man (*Homo-geometrics*?) is still apparent: level hedges and square fields.

My wellington boots squelch into the waterlogged ground along the right-hand flank of the field. Now, away from people, cars, houses and jigsaws, the sense of space intensifies. I imagine myself shrinking as the sheer scale of my surroundings expands. A thought flickers through my mind, a creeping realisation that on the macroscopic level I represent no more than a biological blip in Earth's evolutionary process, as insignificant to the Cosmos as a leaf blown by the wind of time.

I pause near a stile separating two fields. Above me looms the crown of an oak tree. Although there are over eight hundred varieties of oak, only two are native to the British Isles: the Sessile Oak and the English (or Pendunculate) Oak. The latter has a short stemmed leaf and a long stemmed acorn and is likely to have been a planted tree. The sessile oak has a long stemmed leaf and a short stemmed acorn and is more of a genuine woodland tree.

I cross the stile, walk on for forty yards, then look back. There is a single erect trunk, branching out at a certain height into a few thick limbs, which divide further to form the crown. Its silhouette is jagged and unbalanced. The lower branches have been broken, but the crown is rounded some thirty to forty feet above ground level. My own ink-blot impression is of an old stooping figure complete with curved skull cap.

Oak trees do not produce acorns until they are about forty years old and I know, from previous visits, this tree does fruit. Oaks produce a heavy acorn crop at irregular intervals (three to five years) and this main crop is known as the mast crop. Even now there are a few acorns to be found near its base. The squirrels, magpies and wood pigeons must have missed them so far. In olden days pigs used to forage for the acorns and the woodland would be valued by the number of swine it could support.

The mole hills are a few yards away. Those freshly dug are capped by crumbly black soil. The older mounds are smooth and eroded by wind and rain. I fill my bucket and leave it on one side, to collect on the way back. Then I wander over to the small copse.

The oak trees around the perimeter are topped by irregular crowns, not spherical like the tree in the open by the stile. Their trunks are longer before branching occurs and the growth is clumped towards the outward facing side. Individually they look unbalanced, as if they would fall over in a strong wind, but collectively the canopy formed aloft still takes on a neatly curved sweep, even though no tree is neat in itself.

On the fringe of the copse a few clumps of small trees grow. These are almost spherical in shape, indeed quite symmetrical.

If each tree acquired a personality then I guess the oak by the stile would be the wise old man who had room to mature over the years. Those on the circumference of the copse represent inner-city dwellers, each struggling with its neighbour, clambering for existence in a restricted environment. The symmetrical saplings create the most favourable impression. Their form oozes with potential, yet they are so young they are uncertain as to which way to grow and are topped by a profusion of branches thrusting outward in every direction.

Trees and people need room to grow. Crowd them together and their energy is expended by mutual exertion. Stand them in the open and they become lonely figures, hardened and unbalanced by fierce exposure to the elements. When trees grow in smaller clusters they seem contented. Like a family group they grow neither in isolation nor among the multitude, but in a more favourable arrangement.

Walking through the shrubs of hawthorn and holly, I arrive at the clearing in the centre of the copse. A circle of silver birch contains the area like a wooden Stonehenge; their wispy branches dance in the wind, looking fragile and delicate.

Dusk is rapidly approaching. The eastern sky is mauve-grey and the wind increases its chilly breath as night approaches. Behind the silver birch, dominating the skyline, is a huge electricity pylon, part of the same circuit I encounter on my walk to the bus stop. The suspended cables are grouped into six pairs. They resonate in the wind and emit a low mournful drone. The unearthly low frequency groaning and the shadowy surroundings give me a feeling of foreboding. Indeed, it is known low frequency vibration, in a minor key, produces physiological changes, lowering the blood pressure of anyone nearby. The clearing acquires an evil atmosphere. My imagination conjures up a scene of witches cavorting around the perimeter chanting satanic incantations.

Enough of this; I'm off!

On the way out of the thicket I find a holly bush emblazoned with bright red berries. I snap off a twig to take home.

The sense of urgency diminishes as I leave the thicket behind and collect the bucket of soil. Reality re-establishes itself as I head for the stile.

The cold wind blows full into my face and the lone oak tree sheds remnant leaves across my path. The tree leans towards me, bent by the prevailing wind. Beneath the crown of the tree I notice there are two main branches. Interesting: why not three or four? Further up at the twig end, branching can produce five or six twiglets. This may be relevant later, so I make a mental note of these observations and return to my garden.

My last job of the day is to stoke the solid fuel fire to keep us all warm for another night. Then it is time to show Linda the holly branch.

"Linda. Where's your book?" I ask.

She looks at me knowingly and passes me the plant pot.

"No; your book." I see it on the coffee table and crawl over to get it. I thumb the pages until I find a colourful painting of a holly twig.

"Look, Linda." I have left the holly in the kitchen where we can see it above the washing machine. "Holly."

She takes no notice and looks puzzled, so I hurdle the child safety gate across the kitchen doorway and hold the holly twig aloft against the relevant page. Linda crawls over and hoists herself upright against the safety gate. She makes an unexpected grab at the holly. Phew! Just missed.

"Look," I repeat, "here's some holly, just like in your book. Everything about the whole world is in books." An exaggeration perhaps?

Linda squeals for joy as I hand her book back. I pin the holly to the kitchen wall where she can see it, out of reach.

Books are a valuable source of information. I keep the majority of mine in the attic – four shelves full – and I suddenly recall one book is all about trees. Perhaps this will reveal the best shape for a tree, which may throw some light on the goal of my tree of life.

But now, I'm afraid, tree studies will have to be postponed until another day. The attic is not easily accessible and Linda and I are behind on our play for today.

We have got a two-piece jigsaw to sort out now. It is fun solving puzzles, especially without outside help. There is a challenge to finding your own solution to a problem. Books can offer advice and guidance, but always read them critically. Read this one critically, please. Beware the expert and admire the enthusiast. The latter has a mind free of rigid thought; the former propounds a finite subject full of vague terminology.

We have now considered the roots below ground level, a subterranean history of events and encounters that preceded our birth and which we cannot alter, buried by the debris of time. This is our origin, but what is our destiny?

A small oak sapling has a long journey to make before it can reach the 400-year-old splendour of a mature feature tree.

The mature tree below is surrounded by a bench for people to sit upon, to contemplate their feelings; perhaps fall in love, or even get married under its majestic canopy.

To *roughly* estimate the age of a mature tree, measure the girth (circumference) of the trunk 1.5 metres from the ground in millimetres, then divide by 12mm. A 4.8m girth = 400 years.

In imperial units, measure the girth 4½ feet from the ground in inches and divide by 0.47.

Roughly speaking, a mature tree's age is one hundred years per foot of diameter – this is my preferred method of estimating.

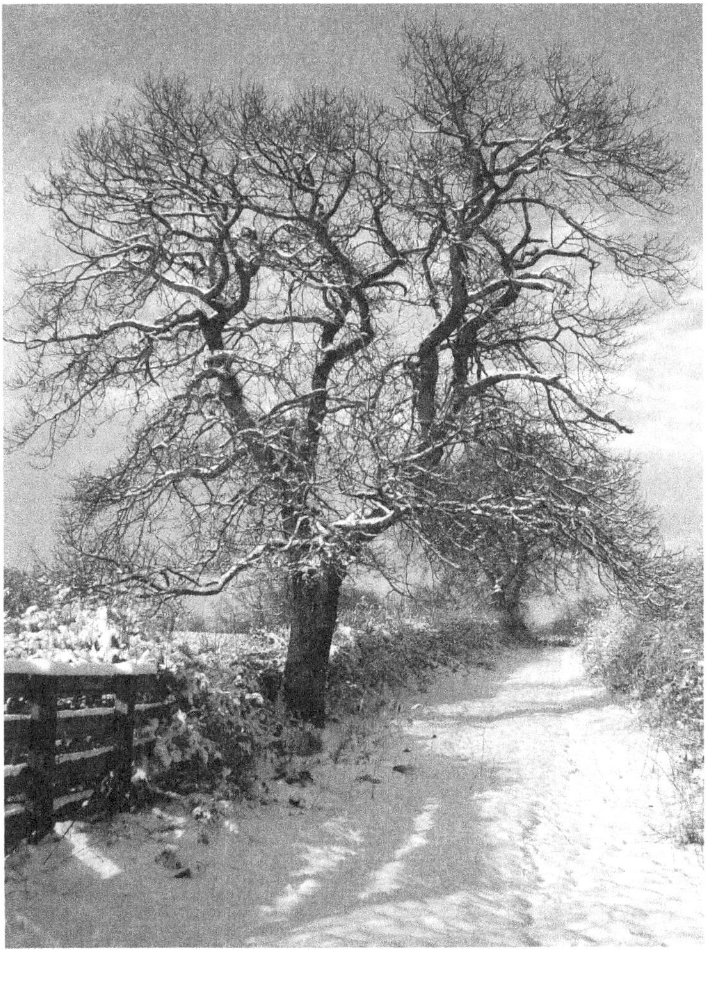

Branch 8

First Birthday

It is Linda's birthday today! It is also December and snow has been falling all night. Nevertheless, I make a gallant effort to wade across the hill to the bus stop and (will miracles never cease?) the bus arrives on time and we reach the factory only five minutes late.

Everything is cocooned in a silky blanket of crystal white snow. I have never seen snow like this before. It has adhered to walls, hedges, trees; even the holes in chain-link fences are filled. It is cold, minus five degrees centigrade, and the roads, including the main roads, are veneered with compact snow which creaks under tyre and crunches under foot.

At dinner time a colleague and I venture out of the factory to the adjacent works' social club. We sip two pints of lager while we socialise. Later, as we make our merry way back, three schoolgirls from the local village bombard us with snowballs and we retaliate with agility and fervour which belies our age and enhances the reputation of lager.

The scenery is superb. The Sun has managed to creep about fifteen degrees above the local horizon and casts long shadows on a glistening background. The trees take on a two-tone appearance; white on the north facing side, black on the sheltered side. Their long, grey shadows pivoting around the huge sundial gnomon trunks; canopies are frosted and still, motionless against a watery, blue sky.

Tranquillity permeates our surroundings. The cold air seems to amplify distant noises yet simultaneously mutes harsher sounds, like car engines.

The forces of nature have buried the earth and reclaimed their supremacy over mankind. The blizzard has created an omnipotent uniformity without bias or favour. Everything has been obliterated: fields, streets, houses, cars, litter and dog poo are all caked with snow. And out of this uniformity the most prominent feature of the rural scenery is the trees.

Autumn was a good time to formulate a few ideas about tree shapes when they were stripped bare. Now the snow has made them a focal point in the winter landscape. They mark where roads are, they outline the fields and sometimes they are clumped together, forming dark islands in a sea of white foam. This is, without doubt, the best time to study the overall form of trees. They stand above the land in stark poses; there is little else to catch the eye. All the leaves are buried, the grass is submerged, the flowers are gone and small, furry creatures are hibernating in secret hollows. It is almost as if nature is helping with my investigation by eliminating all the distractions except for the subject of the experiment – trees. This is an opportunity too good to miss.

Unfortunately there is no time for observing the countryside on that day. Even the usual panoramic views from the bus are obscured by dirty opaque windows in the semi-dark light level. My only chance will be to take my bicycle out on to the lanes one dinner hour.

That evening at home, Linda's birthday cake is still intact. Her grandparents have not been able to visit due to bad road conditions. The main roads are clear; the difficulty is getting to them from the icy white side roads.

Tracy is disappointed. After making all the jellies, cakes and trifle, nobody has turned up to eat them. Yet Linda is not the slightest bit bothered by all this. In fact, since she doesn't know what a birthday is, she is not concerned at all. Two toys arrived by post a few days ago and with our gift there are three new toys for her amusement and we have a good play.

When Linda has gone to bed I hoist down the attic ladder and ascend to the library in the loft to find the book about trees.

There it is; the bright green cover is easy identification. I return to the warm air below and close the attic hatch.

After studying for a little while I find the information I seek: "Trees are the strongest and fittest members of the plant realm," it says and then continues, "They usually have a single erect trunk. This divides to form a crown which may be broadly ovoid to round. This shape provides the tree with the greatest amount of air and light, thus assuring the best conditions for the nourishment of the tree."

Thus the shape of a good tree is round, or more correctly, spherical. So for my analogy with the tree of life, the individual's tree structure should aim to be roundish. This implies symmetry which in this application I will define as being balanced. Therefore we can imagine if the career branch and the family branch are at similar stages (phases) of development, the overall structure will have balance and this will be a favourable condition. If, however, one had, say, a large family branch (many offspring) and stunted career branch (no job), the symmetry is gone and the structure is stressed. It is fairly obvious once it is pointed out and that is what I like about tree theory so far: once the recipient has been told the idea it immediately appears obvious, like all good ideas. However the good tree shape is a somewhat different consideration from the actual goal of a tree. Yet, I feel these two aspects are somehow related and may not be separable.

There is a sharp frost that night and when I head for the bus stop the following morning I find another inch of snow has fallen. I leave earlier than usual and discover convincing evidence to support a fore-mentioned theory of mine – getting up at six-thirty in the morning is not natural. Here, thanks to the snow, is a transient impression recording the fact I am the first person to walk over the hill. Six-thirty is not a natural or normal time to arise, for if it were my footprints would be mingled with many others at peak time of activity.

To be first or last, early or late, are at either end of a distribution curve where the mean is normality.

It is my intention today to make a concerted effort to study real tree structures in these ideal conditions. I know a good tree has a spherical shape and now I have to try to discover the goal of tree structures.

I reach work. Time drags. Eventually dinner-time arrives. After devouring sandwiches and a mug of hot tea I wander over to the bicycle shed. My trusty bicycle (cynics call it my rusty cycle) is chained upright in a metal runner sheltered under a corrugated roof. Here the machine rests, obediently waiting for sunny summer days to transport me in the general direction of a foaming pint of ale – bitter shandy. I imagine the machine is wondering what is happening today as I unlock the chain and reverse its tyres onto fresh untrampled snow. We have never been out in freezing conditions before.

The factory's internal roads have been laboriously freed of ice by snowplough, shovel and rock-salt treatment. But outside the factory gates the roads are still ice-laden and I think it prudent to leave my bicycle chained to a gate while I continue on foot.

When the blizzard first struck, the whole country was brought to a standstill. The snow and ice had trapped the equivalent of three inches of rain. Next it will thaw and there will be floods. In summer a fortnight of dry weather and we enter drought conditions. In the spring and autumn one day of continuous rain brings more floods. Each minor deviation from our normal weather causes chaos. And because the weather normally is temperate we do not make one iota of preparation for the next blemish. I think the weather unwittingly conditions us to a life of moderation. We frown upon extremes, even in our social and political outlook. This is a good situation up to a point and tends to provide a degree of stability in a changing society. But if overdone this easy-going attitude can lead to stagnation by acting as a brake on progress. Any unconventional ideas, like the Earth going around the Sun, or an electrician reforming mankind's hierarchic chaos, meet formidable opposition before the merits of the case are demonstrated. There must always be room for the growth of ideas, otherwise society and its members will become unhealthy.

Broad-leaved trees, like the oak, have three fundamental shapes:

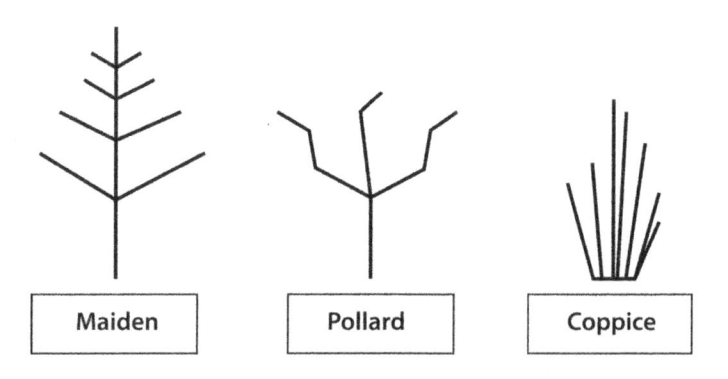

| Maiden | Pollard | Coppice |

The maiden shape is the natural form of the tree when grown in ideal conditions. The main trunk grows into a vertical rod and branches emanate from the trunk from bottom to top.

The second type is termed pollard. Here the main trunk terminates, say, about one-third of the way up the structure and large branches radiate out from this point.

The third type is usually the sole result of man's activities. The tree is cut just above root level and it responds by sending out a cluster of fine shoots, forming a kind of bush. This style is known as coppice and will not concern us further. Only the first and second types are common in the wild hedgerows.

I continue walking along the lane which runs parallel with the factory fence. A wandering Alsatian dog joins me. I pray he is well fed and friendly. His presence means I walk one hundred yards before looking up at the first tree.

Tree species are harder to identify at this time of year. Their fingerprints, the leaves, are not on prominent display, though a few crumpled ones hang on high branches. Yet incredibly, even in these freezing conditions, the trees are forming buds which even a self-confessed amateur like me can identify.

The Alsatian wanders off at a tangent into a short cul-de-sac. With all this snow around he appears to be having trouble finding something revolting enough to have a really good sniff.

I continue along the lane now curving gently to my right. There is a short row of trees ahead, evenly spaced and undoubtedly planted by geometric man. The dark shiny buds reveal their identity; they are young horse chestnuts. Each is about fifteen feet tall and their branches form a well-balanced and symmetrical framework. The lower branches have been sawn off, probably to prevent the local children from clambering into the crowns.

Two joggers pass by. The movement of one runner is accompanied by a rustling noise. I suspect he is wearing a black plastic bin-liner beneath his track suit for extra warmth. I feel a kind of sorrow for them. If only they slowed down and looked around at this glorious scenery I am sure they would feel just as refreshed. Now all they see is the creaking white snow at their feet as they blow, gasp and puff at the icy atmosphere.

Two country houses lie ahead. The first has a slate plaque hung on the gate with "Fox Covert" engraved on it. The second has a wooden off-cut into which "The Yew Trees" has been burnt and two yew trees flank the entrance gates. Next to them is a tall holly tree about twenty feet high, a superb example of undisturbed growth. The tree must be a respectable age, but there are no red berries; the birds will have eaten them all by now.

The Sun shines from a shallow angle above the hedgerow on my left. I reach an open gate and wander a few feet into the field beyond. In the distance the Sun's nuclear globe floats on a pale, silky sky. The horizon is littered with the dark silhouettes of mature oak trees. I scan the horizon and focus on the more prominent trees. Each tree is a true individual and their frameworks seem to be the product of random forces. The only common features I can identify are: first, a tendency for the trees to lean away from the prevailing winds: second, no matter what the shape of the underlying branches, they nearly always produce a semi-circular crown. Very few trees have a balanced skeleton. Most are zigzagged and jagged. Branching initially occurs at varying heights and the pollard style predominates; there are no maiden-shaped trees to be seen.

Two more joggers plod past as I rejoin the lane. I check my watch – there is no need to run. We are so conditioned for running through life we seldom find time to pause and consider which direction we are running.

I am nearly half-way around the circuit and the oak trees line each side of the lane at a fairly regular spacing, the twigs of one mingling with the twigs of the next and so on. I stop beneath the crown of one large oak. Protruding from the sunward side of the trunk, about eight feet above field level, is a big brown fungus. It looks like a giant clam waiting to snap at unsuspecting birds. Young side shoots have grown out of the lower trunk and round woody oak apples decorate the new growth, produced by gall wasp activity the previous spring.

The next oak is choked from ground level to tree top by vigorous clinging common ivy. Out of the hedgerow strange deep-frozen flower husks stand like rows of asterisks on an invisible musical stave. Imprinted in the snow at my feet is a trail of alternating arrows made by the feet of a water hen probably looking for an unfrozen pond. The poor bird will be unlucky today. The lobes of my ears are starting to sting; that means it is still well below freezing.

It is time to circle back along an adjoining lane. There is hardly any colour in the scenery. It is mainly black and white and hazy grey shadows. I pass a farmyard and a sickly sweet smell of fermenting silage hovers in the still air. I wonder if the farmer is having trouble getting the milk to the processing plant. A lot of farmers were cut off, isolated by huge snow drifts, and thousands of gallons of milk have been poured away. In fact it has been so cold lorries and tankers are failing to reach their destinations because their diesel fuel is emulsifying, congealed by the extremely low temperatures.

That is the funny thing about extremes: you have to alter your way of thinking to cope with them. Take last weekend for example. When I brought the milk bottles inside the house, the silver tops had risen an inch above the necks of the bottles on columns of frozen cream. Later, while making a cup of tea, I removed a bottle from the fridge and the milk had returned to its normal level. It suddenly occurred to me the fridge had warmed the milk, melting the column of cream. The idea of a fridge warming things up seemed strange, yet I have heard Eskimos put their milk in the fridge to keep it warm. Extremes can invert normal thinking and this attribute is often employed to test things. Metals are stretched to breaking-point to see their response and new ideas are extrapolated to their limit to see if they too will falter.

Yet another jogger passes by. He is sprightly on his feet and is either a cross-country champion or very keen to impress. Perhaps he will collapse around the corner when I recede from view.

A new smell tickles my nostrils, that of frozen greenery. I look over the hedge and see row after row of frozen vegetables. They are either cabbages or the flattened tops of sprouts. The field looks like a huge vegetable meringue pie. The snowy meringue is glistening and crisp and put on in folded wind-formed layers.

The poor old farmer has tried to diversify, utilising some land for crops rather than all grazing pasture. Perhaps he has tried to avoid the fluctuating rewards of dairy farming, knowing specialisation is only effective with a secure and expanding market, or a monopoly of the product. Unfortunately, even with his eggs in more than one basket, the arctic conditions have damaged both aspects of his production.

The farmer (in fact) is not old. He is middle aged, quite amiable and this weather would have to persist for a long time before he became poor. But we do live in transient times when having one iron in the fire is not advantageous. To me, the maiden-shaped tree with its single erect trunk expresses this mono-dependent life style. Dominated by one single factor, its branches from root to crown are pinned to one central support. There is no major diversity in the structure. If the main trunk becomes damaged or diseased the whole structure is stopped. Fundamentally, the main trunk is the whole structure upon which the leaves and twigs depend.

The tree structure which I find most appealing is the pollard style. Remember, this has large main branches expanding upwards from a short trunk. If this structure is, say, struck by lightning or fractured in a gale, there is a greater chance only one branch will be lost and not the whole tree. A branch may die but growth will continue elsewhere. It creates an impression of resilience which I prefer.

As I approach the factory's perimeter fence the first low rumblings of machinery disturb the static air. Here a green-belt of land has been cleared between the hedgerow and the fence in compliance with a long-standing agreement between the firm and the local council. This strip of land is about forty yards wide and has recently been planted with hundreds of young trees. The saplings help to reduce the harsh visual geometry of the modern factory. Designing factories to merge with the countryside costs too much, so they build in the style of the modern movement: cubes, rectangles and squares, then, to try to blend them into a rural setting, the factories are camouflaged; the sharp edges are blurred by planting trees. If the idea works it is only because we have become accustomed to seeing these contrasting styles in close disharmony. Nearly all architects use trees in their visualisations. There is always a tree bodged on to the final projection. Is this to breathe life into their geometry? Is it to create a feeling of permanence, or a device to add proportion to their composition? Maybe it is a subliminal admission of the ugliness and lifelessness of rigid geometry.

There is one thing striking in young trees. They are a picture of healthy vigorous growth. They are nearly always symmetrical, with a profusion of twiglets branching out in every direction as if probing and examining their surroundings. And I see here a strong analogy between the young tree and a young child also eager to explore his habitat. The young tree is feeling out his environment, the child increasing his knowledge. It is all about expansion: all-round growth. Only then is a stable, healthy and whole organism produced.

There are a great many things that can thwart this goal and often the symmetry is lost by the time the adult emerges. Ideas that used to be flexible become more rigid as we age and can easily turn into a hardwood branch of dogma. The structure is shaped by the prevailing winds that have flexed its branches during the formative years. As the organism reaches maturity the growth rate decreases. Older, unstable structures have difficulty achieving a balanced existence. If the prevailing wind alters direction the structure may topple or be damaged. Eventually the organism ceases to exist.

In the young organism growth is paramount. The concept of no-growth (death) is unknown. As a friend of mine once stated when referring to his youth, "Dying wasn't in the ball game then." No-growth is left for older trees to ponder upon.

The mature symmetrical tree presents the same surface area to the wind no matter which way the wind blows. With no extremities in the structure, the wind's disruptive turbulence is evenly dissipated by the flexibility of the diverse crown. Little damage occurs and the organism has a better chance of long-term survival.

We seem to have travelled a long way from the simple career fork which I originally examined. Now I was truly examining a structure with the potential to meet all the criteria I had previously desired, a structure to replace the rigid pyramid hierarchies that generate so much conflict and division.

I made a list of ten salient points:

1. A tree structure is easily recognisable by old and (more importantly) young alike. Pyramids normally only exist conceptually in the mind, but trees exist all over the place and in real life for everyone to see.

2. A tree is an organic structure closely akin to man. I quote from Carl Sagan's book, *Cosmos*: "Human beings grew up in forests; we have a natural affinity for them." [Good] "There are ... about fifty molecules used for the essential activities of life. And at the very heart of life on Earth we find these molecules to be essentially identical in all plants and animals. An oak tree and I are made of the same stuff."

3. Each part of a tree's structure stems from another part of the same tree in a continuous flow. Pyramids are built with blocks, each an isolated entity, a discrete unit needing a quantum jump to attain the next level. The tree also needs order, but its rules can bend within limits by varying the strategy needed to attain the common goal of the structure.

4. Trees are pliable because of their continuity and this improves feedback and communication. Pyramids are far more rigid; specific rules define the dimensions of each

block. A lot of rules are necessary to maintain each block and these rules can often be right for the block without being good for the pyramid as a whole.

5. Unlike the pyramid, a tree structure is less of an isolated entity. The tree's roots, formed over many years, create a sense of heritage for which it feels responsible. They anchor the tree to the rest of society. The pyramid has shallow foundations. The modern company often exhibits no moral, social nor environmental conscience and lusts only after profit.

6. Tree structures can have a great variety in their formation. Some have two main branches; a few have three or four divisions. Even so, we can still specify for each individual structure an overall shape, a good shape. The outline of a good tree is spherical. At a distance we see trees mainly in two dimensions and would say a good tree shape is round from this viewpoint.

7. A tree structure flexes when exposed to the wind. Prevailing winds shape its growth, eventually forming an efficient structure which causes minimum turbulence. It is evolved for a changing environment above ground while roots provide stability from below. On the other hand, pyramids have good aerodynamic properties that tend to isolate them from external influences. They react slowly to change, relying on their sheer mass to keep themselves in place.

8. The wind does not blow below ground. The past therefore remains buried and is unaltered unless the tree dies. Then the roots die too and the memories contained within them

are lost. The roots decay; their pathways are digested and dispersed into the soil to replenish the earth.

9. The tree's goal is expansion, so too is the pyramid's. But a good tree aims for even, all-round expansion of every branch in all directions, expanding from a small globe to a large spherical crown. There is no known good shape for the abstraction of pyramid hierarchies; therefore its expansion is a hit-and-miss affair.

10. Trees succumb to natural cycles: summer and winter: growth and consolidation. The pyramid perpetually desires growth. It tries to force growth, even in a bad climate.

I collect my bicycle and re-enter the factory gate. The fresh, sharp air and the spectacular scenery have been most refreshing. A lot of ideas have fallen into place. Here was a new structure for man to tend, one offering hope to the millions of drones in the existing model, the powerless people, subservient to the rigid rules of the pyramid hierarchy.

How well I remember the feeling of intoxication that swooned over me when these ideas began to take root. Ha!

On either side of the factory road I see them sitting in offices in front of their rectangular desks. Square blackboards mounted on oblong walls display the white-chalked thoughts that have flickered through their minds. Some look out of the square windows with dull blank expressions etched on their faces. Others spend hours sitting still, interpreting technical jargon printed on standardised A4 paper. Still more are writing dull technical jargon for others to scrutinise.

Their lives are fashioned by mental boxes from the moment they are born. We are bred to fill the boxes so if someone leaves the pyramid the cavity can be neatly filled by a person of the same dimensional ability. Standard, stereotyped people are manufactured to fill the gaps and when they succumb to no-growth the final irony occurs ... real boxes transport them to eternity.

As I ride onward my mind calls out to the people: "Follow me and I will free you from the myth you are preserving. Let me demolish the walls that contain you. Let me unleash the dormant creativity that blind devotion to a dying system has atrophied." But no one is moved, no one can reply and nothing is altered. The pyramid is a very stolid structure.

The line between euphoria and paranoia is a narrow one. I knew I must suppress the former and avoid the latter. A temporary lapse of concentration and I could overlook something which might destroy tree theory in its infancy. Thus the feeling of contentment was a very temporary affair. The theory had to be applied to practical situations to see if it could survive.

All these thoughts were on my mind as the afternoon expired and the bus arrived to take me home. An old beer can was rolling around on the bus floor that day. The driver reacted by stopping the bus, opening the doors and slinging the offending tin into a bus shelter. Meanwhile I was making the mental adjustment from factory clone to family man and today this process was laboured. A proliferation of questions flickered through my mind. Each would have to be answered for tree theory to survive a closer examination.

I present here two recent images of the maiden and pollard style trees.
First the maiden tree shape below ...

It is a wonder it has managed to survive in the middle of the field. At the bottom you can see the grazing lines of past generations, perhaps cattle. This tree has grown to attain the ovoid or spherical shape mentioned in tree books, allowing sunlight to replenish it from dawn till dusk.

Meanwhile the pollard-shaped tree above has had a harder life. Sometime in the past (a few hundred years ago) the main shoot was broken and it branched out into new areas. It is a less compact tree, yet still trying to attain its natural shape by reaching out with balanced symmetry.

Branch 9

The Blizzard

I have little faith in optimistic weather forecasters. On Thursday night the weatherman stated quite plainly snow was on the way but the front carrying the blizzard would not reach northern parts. In work next day at around ten o'clock, and at a latitude well north of the southern parts, the blizzard commenced. The temperature rose to four degrees below freezing and the land, still white from the last session, was engulfed again.

Wet salted roads became white and at two-thirty in the afternoon an announcement wafted out of the works' public address system: "Attention, please. Attention, please. Due to the bad weather, employees may leave at three o'clock. Those travelling on public transport will not be affected. Thank you."

And thank you, too!

Fortunately I wasn't on the bus that day. I had arranged a lift with a colleague who passed by my house on his way home. We left at three o'clock prompt. My foreman wished me the best of luck and I returned the gesture.

Jim, my chauffeur for the day, had dug his car out of the car park and off we slid on mushy brown snow-laden roads with the windscreen freezing over, the washer jets blocked by ice and the blizzard raging outside. The car ploughed on. In an attempt to raise morale, I said, "More power, Jim. The storm is at its height!" His response to this paraphrase will not be found in this work; in fact, portions of it are not yet in the Oxford English Dictionary.

I am glad I'm not driving. The main road is churned into a soggy pulpy mish-mash of slush. The dual carriageway is limited to one lane if you are sensible. Insensible drivers still exhibit a need to overtake in the outside lane where there are no tracks to follow.

Ahead is a natural gradient, a long gradual hill climb rising about two hundred feet. There was a blockage here during the last spell of bad weather when two large container lorries failed to negotiate the slope and, since one was overtaking the other at the time, they formed an effective barricade to those travelling in their wake.

The traffic continued to move in a long, snaking procession and Jim comments there are a lot of people on the piste today. Judging by the way some are driving he is *not* mistaken. You can tell the well-off and affluent; speed is inversely proportional to the amount still owed on the vehicle.

Well, we make it. Phew! I clamber out and watch Jim's car skid away into the white oblivion. By the time I have trampled a path to my bungalow I am encrusted with snow. The damned stuff has filled the inside of my hood and lodged around my ears.

"You're early!" Tracy exclaims. We are both relieved I am home and in one piece. Linda appears and throws a yellow plastic duck in my direction and we all laugh.

The fire is glowing red and there is enough coal to keep us warm overnight, so we sit out the storm until the following morning.

Why is it so satisfying to sit around a warm fire while the weather rages outside? Surely it is more than the obvious physical advantage. Perhaps it is a smug feeling of mastery over the elements, being in control of your own mini environment, no matter what random event Mother Nature conjures up next. You shelter inside your own portion of thermal stability in the midst of an unpredictable atmosphere. The largest computer in the country tries to compile tomorrow's weather, an optimistic weatherman understates the result – yet no matter; a coal fire restores equilibrium. No one can predict tomorrow. There are just too many variables involved for an accurate result. Only general trends are sometimes noticeable.

The next morning Linda is first to wake up and by six o'clock she has made enough noise to be allowed into the big bed between Mum and Dad. We snuggle down until seven-thirty, then Tracy turns the light on. This is Linda's cue to start messing about–playing.

"Sit down, Linda."

"Linda, don't put my watch over the –" Too late. It falls behind the headboard and clanks on to the central heating pipes below.

"Okay, down you go." I lower her to the floor. "Quick, Tracy. Grab the clock. She's coming."

After about fifteen minutes, Linda has given me quite a collection of things: a comb, a slipper, both of my work shoes, the top cover from an electronics magazine (where's the rest of it?), my torch, an envelope and a sock which was airing on a radiator.

"Yes. Ta, Linda," I drawl.

I get up and dress, have a wash and begin a tour of inspection around the bungalow – from inside of course. About one foot of snow has fallen, but the wind has sculptured it into uneven distributions. The side of the bungalow is relatively clear, the cement drive just visible. The back garden paths have been filled in again and the path to the coal bunker is buried. The garage door is submerged under a big drift of snow.

Now for the front garden–it's bad. To reach the gate I will have to dig out a wedge-shaped cross-section of snow three feet by the fence tapering to one foot deep near the lawn. This would take all day probably and then there is still one hundred yards to go before reaching the main road, which should be passable. This leaves little option but to cancel the shopping trip and leave the car in the garage.

Below is a scanned image of "The Blizzard" taken three decades ago.

This is the last Saturday before Christmas and, as usual, there are still things to buy. Looks like a last-minute panic again.

Christmas is a very expensive time of year. It shouldn't be, but the Biblical message has been superseded by the capitalists' desire to sell you something, even in the midst of a recession. Cynicism aside, it is nice to allow oneself a bit of indulgence once in a while. The problem is to do it when only one wage comes into the house and a basic wage at that. There is talk of redundancies at work and overtime (a traditional source of vital extra income) is not forthcoming. No "Turkey Sunday" this year. Also, on the debit side, we have a record mortgage rate (the highest ever) and rapidly escalating fuel costs. The price of coal has doubled in five years.

Of course, our solution is to spend less. This is what the government is doing and so do we. We have to, but the bank account nears depletion and our savings are measured in shoe sizes. I cannot imagine the distress and the strain this must place on those not earning a so-called "average" wage. Perhaps I will discover these things soon.

After breakfast I stoke the fire and don my full winter clothing ready to tackle the drive.

Below the kitchen window, near the side door, a hand-held snowplough lies wedged below the windowsill. Using this, I forge a path to the garage where the car lies covered with dusty white snow. Nearly everything is topped with fluffy snow, blown in through the corrugated eaves. Paint tins, boxes of screws, an old car radio and the tools above my workbench are all frosted over.

I collect the shovel and start clearing the drive. There is activity at nearly every household along the street. The sky is clear; the air is fresh and filled with the sound of spades clattering and scraping cold cement. Methodically, I work my way towards the road, pausing frequently to observe the unusual setting. Others dig furiously, as if their very lives depend upon it, ironically risking a heart attack in the process. Perhaps the scalloped snow drifts, the gentle silky curves and the glistening panorama are just too much for them to absorb. Eagerly, they try to return their immediate surroundings to their usual shape, which they can relate to better.

It appears vital to dig the car out. How dependent we have become upon this machine. At the moment the car might as well be stuck in the garage, or on the drive, as stuck on the obliterated street. Last year the local farmer helped matters by using his tractor to crumple tracks in the snow around the estate's roads. Patience; he may turn up again later on.

It takes two hours to clear the lower drive and all the way to the road through one wedge-shaped drift. I'll leave the deep snow until tomorrow.

When I go inside again my clothes are soaking with sweat. I have a full change of kit and a mug of hot tea.

In the afternoon I take Linda out to look at the snow. We do not have a sledge yet, but she does possess a small tricycle with an "L" plate mounted beneath the seat. She sits firmly on the seat gripping the handle bars while I crouch over and push her along. The Sun is making an effort to warm things up, but it is still below freezing. We stop next to the three-foot snow drift, which I started to cut through in the morning, and Linda paws at the funny white powdery stuff with her green mitten. Half an hour later my back is aching and my feet are cold, but Linda is still waving to the neighbours and gurgling with joy. A few minutes more, then we go inside for more tea and biscuits.

Later that night, when all the chores are finished, Tracy reminds me about the factory's Christmas outing. Next week my colleagues and their wives are going on our annual treat to club land. The question is: can we afford it? On this occasion my meagre savings provide the answer, money accrued during pre-recession days.

It has been apparent for some months now Tracy and I are only just financially surviving. Present government policies seem to enhance the recession, not ease it. To use economic parlance, we are in "real terms" getting poorer. I am hoping this is a medium-term problem and I am wondering what to do about it when I suddenly realise this is a good time to try out a little bit of tree theory.

All theories are instruments which we apply to reality to help solve problems. They offer guidance out of the confusion of choice by focussing our attention and concentrating our wandering thoughts.

Earlier I drew a career fork when faced with a decision about my working life. Now I gathered pen and notebook to create a financial branch using the same simple technique. Expansion of this branch was now desirable. What options were open to me? I make a sketch.

```
        *
    ↗ ┌──────────────────────┐
   ╱ ↗│ Cut back on expenditure │
  ╱ ╱ └──────────────────────┘
 ╱ ╱ ↗┌──────────────────┐
╱ ╱ ╱ │ Promotion at work │
 ╱ ╱  └──────────────────┘
  ╱ ↗┌──────────────────────┐
   ╱ │ Part time / evening job │
    ╲└──────────────────────┘
```

┌─────────────────────────┐
│   **Financial Branch**   │
└─────────────────────────┘

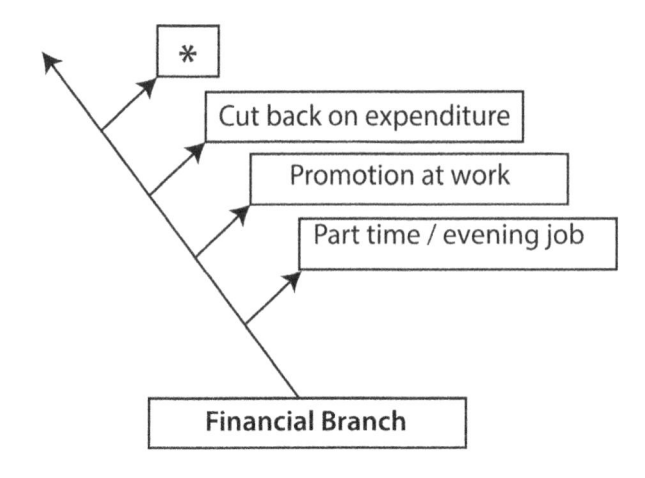

Each bud is an idea, a potential path for my financial growth, and we must remember it is not an isolated appendage; other branches intertwine with it, like the Career Fork and Family Fork of earlier. As another example, attending an evening class (for further education) may expand both the career and (hence) the financial branch. When viewing the overall tree there is a position where the branches seem to meet, or interact. In tree theory "lateral drifting", mentioned earlier, is not prevalent as it implies no forward motion. Yet time moves on around us and we age regardless of our indecision.

Let us examine the financial options.

First bud (or idea) is a part-time job. There are still a few of these about if one is not too choosy. This could only be considered as a temporary solution without any real long-term benefits. It would be physically exhausting and create extra stress on the family branch of my tree.

Promotion is the next thought. This seems a very rational solution, but if a redundancy situation arises it complicates matters, as we may see later. Other questions come to mind: is seeking promotion for financial reasons alone an ethically good objective? Shouldn't promotion be for those who only wish to grow and expand their knowledge? Does the financial carrot attract the right kind of person? My answers are no, yes, and no respectively. But perhaps these are ideal answers imposed upon a far from idyllic world.

On we travel along the branch and the next bud to contemplate is a reduction of expenditure. In terms of tree theory this is equivalent to a bit of pruning. The difficulty is in knowing when to stop cutting, hopefully before the tree becomes so poor that it feels like it is hardly worth living. Pruning is anti-expansion but is not in itself evil, providing too much damage is avoided.

By Western standards no one could accuse Tracy or me of extravagant living. That is, nearly everything we own has a functional purpose born out of necessity. I even class our small car as a Western necessity and sometimes an evil one at that. We do not indulge in the social vices; we don't smoke and we are not out drinking every night. If we do spend an evening out, it is a rare treat. Tracy makes a lot of our clothing and I repair everything that breaks down, within reason. Both aspects are very economical modes to exist. Exist: at the moment that is all we do. All the money I earn pays bills, nothing is saved. There are few areas left to prune.

I was going to tell you, "I'm not complaining," but obviously I am. Our family group could be described as comfortable and I will accept that. I am just trying to explain pruning is more applicable to fat trees, while ours is a moderately lean one to begin with.

Nevertheless, we are forced to make further economies. To save petrol we cut down on car journeys and we omit the bottle of weekend wine. I cut down on my impulse buying of electronic magazines and Tracy survives without an inflation-proof household budget; her housekeeping money is the same as last year.

"What are you drawing?" Tracy asks.

"Oh, I'm working out my finances."
"That shouldn't take long." Pause. "Well?"
"What?"
"Well, what's the solution?"
I consult my notes.
"Well," I begin, "I can either get a part-time job ..."
"Ha, it would kill you," Tracy mocks.
"... we could cancel the papers and sell the T.V."
"You won't!"
"... or I get promoted."
More derision greets this pronouncement.
"Robbing a bank would be easier!" says Tracy, but her remark hits me like a brick wall might, because I see this offhand quip in a different light.

You see, at present tree theory would offer no objections to this solution. Robbing a bank would indeed solve my financial difficulties and also be consistent with expansion. So, on the end of the branch where the asterisk now resides, I write "Rob a bank". This anomaly will have to be considered in detail.

A rule is negated if only one event happens which is contrary to its pronouncement. Robbing a bank was consistent with the theory of financial expansion, yet contrary to existing normal behaviour. This dichotomy could ruin tree theory and now I started to envisage other difficulties a goal of outright expansion would produce.

Say we have a branch of an individual's tree of life and this person had to decide how many children to have in his family. With the present goal being unqualified expansion, the answer would be ... as many as possible. Just keep banging them out, diversifying and populating all over the place. Yet it is widely believed birth control, hence limiting the number of our species, is one practical way of ridding the world of the poverty that two-thirds of mankind now suffers. Once again the expansion theory stumbles. It is brought into disrepute; even though there is no doubt that some expansion is necessary for growth. The alternative, stagnation, must be avoided at all costs.

How much expansion can we allow? What limit is good for the tree and what level bad?

The answer was not forthcoming. I was asking questions which had puzzled thinking people with greater minds than mine for centuries. I decided not to force the issue but to consider it at a leisurely pace.

Theories evolve through three stages: first is the desire to solve a problem accompanied by the conscious, or unconscious, gathering of data and ideas. In my case there was a need to understand the nature of the conflicts I encountered. Second, a sudden insight occurs when the isolated entities of thought are drawn together and fused into a single homogenous concept. My inspiration was that tree structures were better suited to guide man's development than the existing pyramid hierarchies. Ecstasy sometimes accompanies this phase.

In the third stage the theory is tested and, usually, one or two anomalies emerge. If there are basic flaws in the theory it must be rejected. However, if the discrepancies are only due to errors in the test procedure, hopefully any impasse can be overcome. The third stage tends to dissipate earlier euphoria and gives way to a more pragmatic approach; a sobering phase.

Theories only have substance when they can be shown to have some beneficial practical value. And later I hope to enlarge upon the benefits of tree-structured thinking. But first I have to clarify the amount of expansion suitable for healthy growth.

I soon realise part of the problem was in my method of analysis, for I had treated the financial branch as an isolated entity, amputated from the rest of the whole darn tree. Here I was preaching continuity, yet guilty of discrete pigeon-hole thinking.

We spend years being taught how to solve problems by breaking them down, bit by bit, until each individual element is solvable. However, solving each element may still not unravel the whole. Our heads cannot consider whole problems in one instant; they have to operate gradually in stages, each a progression from the last step: like chapters in a book; like buds along a branch. As we climb the branch each bud represents an idea or an option, but while we are studying the buds closely we must also remember the part they play in the tree as a whole. If we permit one bud to expand to infinity the whole tree is unbalanced. The other branches are drained of energy and they wither. Hence unlimited expansion on one branch, or *of* one branch, detracts from the healthy all-round growth of the tree.

The chances of successfully achieving a bank robbery are small and the thought of a lengthy jail sentence is a good deterrent to the average person. Capture would curtail other areas of one's life: family affairs, career, hobbies, sporting activities – all gone. So failure means all is lost. But what if the robbery is successful? Amazingly enough, tree theory's denial of extremes means this would also lead to personal failure. In the latter case the culprit is transformed from pauper to prince, usually overnight and at the weekend. The sudden increase in wealth has so many damaging side effects his life is dramatically changed, I propose, for the worse.

Perhaps you disagree and may say, "Ah! But what about a Great Train Robber? Is he not happy in Brazil?" Again my reply is, I think not. Ostensibly he's having a good time, yet still yearns for a pardon and a return to Britain. Nevertheless, let us assume, for argument's sake, the money he stole has had no adverse effect upon his life. He lives contentedly in Rio, interrupted by the occasional kidnap attempt. He then becomes an exception to the tree theory idea of no extremities.

Good.

Good? An exception to a rule is good!

Remember pyramid teaching advises one to ignore all anomalies which detract from the idea of the perfect rule. It preaches the dogma of no alternatives, which is seldom the case and suppresses any apparent dissent.

Tree systems do not have a doctrinaire outlook. What we know as laws of science are merely statistical truths, still in the process of refinement. Tree systems are, likewise, open-ended hierarchies (not closed systems) unblinkered by creed. They consider good rules to have few anomalies, while bad rules have many exceptions and need to be questioned and examined – exposed not sheltered.

Each disparity creates a debate which is usually more interesting than the rest of the problem under investigation. Tree theory recognises nothing is perfect. It is more about consensus and probabilities rather than the arrogant certainty and deterministic attitude pyramid people espouse.

So an extreme in a system will probably unbalance it in a detrimental way. Perhaps I could demonstrate the adverse effect of over-expansion by an example with no criminal element involved, as the latter might cloud the issue. But time is rolling on and I start to tire. Ideas float around unresolved in my mind.

I settle down for supper. The television is turned on just in time to see the weatherman predict another severe frost tonight and a cold sunny day tomorrow.

The next programme shows highlights of two of today's football league matches. There are a lot of goals and I find it quite entertaining. The only disagreeable part is where the experts express their views on controversial incidents. As usual, none of the experts agrees and the viewer is left confused; more confused than if he'd made up his own mind in the first place. We are always being guided and cajoled in everything from politics to sport.

My eyes start to feel heavy and my limbs wallow in gentle warmth, a remnant of the exertion of snow-clearing. I pour a pint of lager and settle back to watch the second match.

The appalling weather has devastated the league programme so the pools panel have met (in a London hotel) to judge the likely outcome of postponed matches. Their deliberations have created a jackpot situation and telegram claims are requested for coupons with twenty-four points total. They list the numbers of the teams that had score-draws and some look familiar. I will check my coupon tomorrow. Now it is time for bed.

The cool sheets soon mellow and warm. Events of the day flicker through my mind in a random muddled order. A welter of ideas crosses between conscious and unconscious levels of thought, searching for an ordered position in my memory. I drift in a twilight zone between reality and dream, where fact and fantasy blend into a distorted vision of one's own desires and anxieties.

The next thing I sense is morning and Linda is bringing me the newspaper.

"Good morning, Daddy," she says, then turns and fades away.

High above my head is a tall branch covered not with leaves but with football coupons. At the tip of the branch a coupon becomes detached. It floats down through the bedroom ceiling and lands gently on the newspaper. I've won! I've won! My numbers match those selected by the pools panel and they are here to congratulate me in person. Their flattened faces press against the bedroom window, smirking at my good fortune.

I'm rich! The tall branch above me expands beyond view to infinity and the tree it is on falls over.

"Tracy, we're rich!" But Tracy is not there. I wonder where you are, Tracy. Tracy?

The bungalow seems different today. Everything has altered. I jump out of bed. The newspaper falls on to the floor with a loud bang. I look down. The headline reads: "POOLS MILLIONAIRE DIVORCED. Ex-wife stays with mother." I rush to Linda's room. She's gone, too. A coupon lies folded in her cot. I read it. It is a ransom demand for her return. She has been abducted; more anxiety.

I open a high window. High fencing, topped with barbed wire, surrounds the garden; keeping a large crowd at bay. A football crowd perhaps or just mankind's needy and greedy? The pools panel snarl at me: "You're rich!" they shout mockingly.

"How do I get to work?" I ask feebly, then the bus arrives, towed by reindeer. The wheels are gone, replaced by skis. It looks like a huge red sledge. The driver seems relaxed. He has a white beard and laughs, "Ho, ho, ho," as we float upwards high above the pylons. The wires guide us to my factory. Below is a long thin line, a landing strip, blood red against the white wastes. Trees line each side of the crimson carpet upon which the bus-sledge lands. Nearby a man is spraying green daisies with white paint, so they conform to the rest of the scene. In summer they spray them chemical-green to blend in with the grass.

My foreman greets me, "Best of luck," he says, then adds, "Five minutes late again."

My workmates are lined up like two teams at a Wembley Cup Final. The foreman steps forward to reintroduce me to each one in turn.

Arthur: "Lend us a fiver, guv'nor."

Steve: "Can you settle this bill?"

Bert: "I've taken your tools. You won't need them now."

Here is my chargehand. "Goodbye," he says. Goodbye? Where is he going? Wait a minute, he's not – I am! I cry out as two security men drag me away. "But wait," I protest, "I don't want to leave! I want to work. I want to be the best electrician in the world, but you won't let me."

"We're sorry," they reply. "You're too rich to work like normal people do. Off you go now, like a good little rich extremist."

Now I hear Linda crying clearly, as if she is close. I jump out of bed, my bed, and arise from my dream. I fumble for my torch, then head for Linda's bedroom.

She is crying, but safe. I wonder what her dream was all about.

"Are you two okay?" Tracy calls out.

"Yes," I whisper. Linda falls back against her pillow. "We've been dreaming." Perhaps our unconscious minds reached out to each other through the cold night.

I totter back to a warm bed. My feet are cold. That's all I feel.

Next morning I check the real football coupon results in the morning newspaper.

"Did we win?" Tracy asks eagerly, seeing a look of contentment on my face.

"No." I reply. "We were lucky."

"Lucky?"

"Yes. We lost."

Her response will not be found in this work.

It is with some irony I have to report current events of the 21st century mirror those of the preceding 20th century. When I began reviewing the prior branch of thought, a modern blizzard swept in to deposit two days' worth of wet snow.

The North Atlantic Jetstream remains resolutely south of Britain and cold Arctic air has met a moisture-carrying warm front blown in from the Bay of Biscay. We face the coldest March for sixty years.

The driveway is buried under snow and it takes two days to dig out again.

Note the new tree to the left of the gate. Linda planted a conker when she was in the Brownies around twenty six years ago. Its expanding trunk is starting to move the gate post, but I haven't got the heart to chop it down. The tree trunk is six inches diameter, which would mean 50 years old by my "one foot diameter per hundred years" estimate for a mature tree. So I think this rule of thumb is not applicable to rapidly growing young trees. Indeed, if one does an internet search there are even correction factors to apply if a tree is in the open, or in the middle of a wood.

There are other similarities, too. When I wrote the original "Blizzard" text the country was in deep economic recession. Thirty years later, we are again in recession and this time it's the *deepest* recession economists have ever seen. Not good.

Finances are strained for all but the rich, so no change there either. Once again the people suffer from a recession not of their immediate making. Indeed, this is a recurrent theme in Britain and globally.

Farmers destroy hedgerows and their monoculture decimates habitat for wildlife. Next they get a subsidy from the people of the EU to set aside land to help wildlife. Globally the emitters of greenhouse gases are damaging the climate. China burns forty thousand tons of coal an hour. But again the bill for the damage will not be paid by those who caused the problems. It's always the innocent people who pay.

Materially, we have come a long way in thirty years. The original "Blizzard" photograph was taken using a chemical process on 35mm film from which negatives and prints were made. Nowadays it's all done digitally from camera to PC to memory-stick to laptop and imported into text: marvellous.

But socially and economically we haven't got much better at all. Economically the majority are worse off. As for society, it's in bits.

The Meteorological Office weather computers can now attain 100 trillion calculations per second! Now a four day forecast is as accurate as a one day forecast thirty years ago, so good progress there.

I still have cold feet.

Branch 10

Galum's First Christmas

On Sunday afternoon Tracy and I take down Linda's birthday cards and pin up the Christmas decorations. Five days to go.

I forgot to mention that Linda can now toddle about unaided quite well. We have to be careful when we turn around in case she has crept up behind. She follows everywhere, watching intently and pointing and babbling at each decoration we touch. Perhaps she thinks our activities are a prelude to the second phase of her birthday. It will be a few more years before she understands the decorations commemorate another baby's birthday, one of the most famous babies ever born.

Earlier this morning I finished clearing the drive of snow and was able to get the car out of the garage. I managed to drive over the hill, on firm compact snow, to our local garden centre. Here I selected and bought a five-foot Christmas tree, which fitted snugly into the car through the open hatchback. Then I drove home again and parked on the clear driveway.

Linda watches through her bedroom window while I wrestle the tree out of the car and into the garage. Here, beneath an up-and-over door and in view of Linda, I wedge the tree into an old bucket using two house bricks. To try to keep the tree fresh I add a few pints of water. Next I attempt to spray it with an anti-desiccant vapour. This exercise turns out to be a total failure. I could not comply with the instructions on the can: "DO NOT SPRAY IN A CONFINED SPACE" i.e. inside and "DO NOT SPRAY AT TEMPERATURES BELOW 55°F" i.e. outside. Instead of getting a fine spray of mist I get a long stream of foam squirted across the garage floor. Messy; too cold. The rules and reality are not in harmony.

There are many occasions when I pause, step back from a job and feel that way. Although my actions at the time seem quite acceptable and conventional, they also seem totally illogical. Christmas trees are a good example. We chop the roots off then try to water the stump and spray the needles with chemicals so they do not die so quickly. Why don't they sell them with the roots left on? Perhaps people dislike a clump of soil on their trees; or perhaps they might use their trees again next year and the year after and shatter the foundations of Christmas capitalism. Either way it seems a shame to kill a perfectly healthy seven-year-old tree just for three weeks usage as a decoration.

I take the tree inside and place it on the wall shelf to the right of the fire. There it stands protected from Linda's grasp by a cunning arrangement of fireguard and couch which has to be seen to be believed. It is the electrician's prerogative to fit the Christmas tree lights, after which Tracy adorns the branches with tinsel, crackers, baubles and chocolate novelties.

Along the walls loops of glittering string await the arrival of the expected Christmas cards. Balloons festoon from above the door lintels, high enough to avoid the opening of doors.

Linda plays gleefully with a spare balloon.

"Say bal ... loon, Linda."

"Ba ... ooo," she says.

"Hurray!" we say.

I can see by the way Linda is squashing and squeezing the balloon she is unaware they go "BANG" when burst. She sees Tracy and me cringing, but cannot understand why and continues to scrape her fingers over the taut rubber.

After tea Linda and I sit on the carpet and play games. The first game is tower building. She is fairly adept at manipulating objects now. Placing one block on top of another presents little difficulty after a brief demonstration and plenty of verbal encouragement. We take turns at building: my go first, then

Linda, then my turn and Linda again. Four bricks high is the record before Linda knocks down the tower and giggles.

We play with a gaily coloured plastic ball next. I throw it to Linda, she grabs it between both hands and carries it back to me. Sometimes she makes a pushing motion and ball rolls my way. "Good girl," she is told. Sometimes the ball drops and rolls away: never mind. When she fails to retrieve the ball it's time for the next game.

Linda has a cuddly toy dog named Mickey. The real live version is a small white Scots terrier that lives with one of Tracy's aunts. This is the game: I have to sit on the floor with my back to the couch and my knees arched up in the air. The triangular tunnel beneath my legs becomes Mickey's kennel. Mickey hides in his kennel and Linda crawls around to see where he is concealed. When she gets close enough, I make Mickey jump out and waggle his nose on her fingers. Linda giggles and laughs and generally gets overexcited, then Mickey disappears for another go.

At six-thirty Tracy prepares Linda for bed. Meanwhile I boil milk and change the solution which sterilizes her feeding bottles and teats. As a rule I feed Linda in the evening, then, if she hasn't fallen asleep, we look at some books.

While adult books present ideas or situations in an orderly way (hopefully), children's books for one-year-olds just present objects; unconnected, unrelated objects. Let us examine *Baby's First Book* with Linda. She is sitting by my side on the couch cradled in my left arm. Her eyelids are heavy and she sucks hard on her dummy. The conversation runs like this ...

"Open the book then. Go on, open it. Good girl. What's this, Linda?" No response, she is too tired tonight. I continue turning pages and naming objects as follows: apple, spoon, dish, brush, comb, tree, watch, dolly, teddy, dog, cat, balloon and buttons.

"Point to the red button, Linda." Her tiny digit lands in the general area of the red button. She has memorised the response to my question, but does not yet understand the concept of colours.

I continue through the pages: mitten, ball, banana, chair, toothbrush, keys; all fascinating objects for one so young. If you fancy an interesting exercise, try writing a paragraph containing all the objects in Linda's first book. You must mention each object in the same order as in the book. It sounds tricky, but there are many ways to achieve this when you know the association between objects. For example, you could say a dictionary contains ball and banana on the same page. Or, if you tread on a ball or a banana you may twist your ankle. In our brains are a multitude of diverse pathways to link differing objects. Imagine a huge telephone exchange directing information and ideas to various destinations, each one a memory cell, or a group of cells. There are one hundred billion memory cells (neurons) waiting to be filled; connected by one hundred trillion pathways, all aching to be activated. The "First Book" is the beginning of memory loading for Linda.

When Linda is packed off to bed, memories of her first book are still floating around in her small brain, searching for something with which to associate. She knows that Mickey is a dog, telephones go "ding ding" and she knows what a teddy and a balloon look like. But there are some items, like keys, which do not readily fall into a settled location. When this occurs the "keys" mentally swirl around her head searching for a favourable area of memory. When she sleeps her brain creates imaginary scenarios to test-fit ideas. Unresolved ideas create dreams.

Many things we do seem perfectly normal, yet when viewed from a different frame of reference are logically absurd. The pathways in our minds have been conditioned to approach certain problems from specific directions. Consequently, our traditions sometimes overwhelm their humble origins and a good example of this facet is the Christmas festive season. Linda is too young to consider Christmas activities absurd. Nearly everything she sees is, at first sight, puzzling, and Christmas is just one more experience to enjoy or ignore. In years to come she will be so accustomed to the Christmas ritual that she will not even consider questioning it to any degree of depth. In a sense, our ability to independently judge situations, or free ourselves from ill-conceived prejudices, is bred out of us unknowingly. The child's innocence is blemished at an early age.

I wonder how an adult life-form from an alien world would judge our activities.

I wonder …

"How was the transport, Galum?" I ask my guest from the planet Trocite. He (or to be more pedantic, it) is a cosmosocial scientist on an exchange visit. He switches on his Posdic unitranslator (made of plastic in Japan) using his third tentacle.

"Terrible," the translator says. "Five light years late. But it will be worthwhile to see, at first tentacle, the unique "Earthiosocial" activity called Christmas. Could you explain?"

"Christmas?" I ponder. "It's a simple celebration really, when Earthlings remember the birth of the Son of God."

"How do you spell that?"

"Christmas?"

"No. God."

I tell him.

"Anyway," I continue, "we decorate our houses with coloured paper, balloons …"

"What is a balloon?" he interrupts.

"Er … a … thin rubber … well, sort of, bag, filled with air under pressure and tied in a knot at one end."

"Aha!" exclaims Galum's translator, emphasis circuitry on full volume. "A BA … OOO! We use them on Trocite too." He winks his only eye by raising the lower lid upward to his foreskull. "Then what?"

"Well, next we cut down a tree, spray it, put it in a bucket of water – so it doesn't die too quickly–and cover it with electric lights, tinsel and chocolate novelties."

"Is this a sacrificial ritual?"

"Er...I don't think so. I'm not sure where the idea originated. We've always done it since I was a lad. It's just traditional."

"Why do you detruncate the harmless carbon based life-form named tree?"

This puzzles me too. No one has ever asked these simple questions before. The telephone exchange inside my brain goes haywire, but no answer comes. There is no reasonable answer available for association. Feebly I say, "Tracy prefers a real tree. I like artificial trees myself."

Galum shuffles uneasily on his claw.

"What else happens?" his Posdic says.

I think; hard.

"We buy and exchange Christmas cards and presents."

"Ah. Interesting. Did the son of ..." (Galum consults his notes) "...um, God, did he start this exchange-of-goods custom?"

"In a way, yes," I reply.

"Then what happened to him? How did Earthkind reward him?"

"I speak slowly and clearly into the translator." You are not going to believe this, Galum ..."

In the above, I chose Christmas as an example of the absurdity within familiar events only because it is close at hand. It could have been any one of numerous everyday human activities, which when analysed appear illogical over a broader spectrum, from not letting people on a half empty bus to, say, the butter mountains and milk lakes of Europe while people starve elsewhere.

Meanwhile, with our feet firmly back on Earth, it is Christmas Eve. Last night Tracy and I went on a works' outing to club-land and Linda stayed with Tracy's mum. Today I am on leave and this morning we are going to call on the parents and collect Linda.

A slow thaw weakens winter's icy grip.

On the trip we exchange Christmas presents, to be opened tomorrow (of course). We have dinner with Tracy's mum and leave about one-thirty in the afternoon. I want to get home safely before the Christmas revellers are ejected from the pubs and office parties on to the main highways.

For Linda, aged one, Christmas Eve is just another night. She drinks her milk and goes to bed at the usual time.

Tracy packs Linda's parcels into a pillowcase, then puts our presents beneath the Christmas tree. We are excited! Linda sleeps on …

Happy Christmas everyone!

Cheers!

"Linda, it's Christmas!"

I switch the tree lights on.

Tracy and Linda unwrap presents while I take a few photographs (you only live these moments once), make a cup of tea and attend to the fire. Then it is my turn to unwrap presents while Tracy freshens up. Soon there is a pile of toys on one side of the room and a mound of paper and boxes on the other side.

Linda "helps" me to unwrap my presents and gives Tracy a hand, too. The cardboard boxes fascinate her.

No jobs for Dad today. Christmas is my day off. If anything breaks today it must wait until next week to be fixed. I am too busy playing with Linda's toys! There is a toy car, a doll, a pull-along train, a duck whose head waggles as he rolls along and a battery-driven plane which plays tunes from four plastic discs.

Tracy and I bought her a shape sorter, full of rigid geometric shapes and made entirely of wood.

My dad has filled a cardboard box full of novelties and wrapping paper for Linda to rummage in, but she seems overwhelmed by all the surprises within; too young yet perhaps. My dad has made Linda a few special jigsaws as well.

We all eat breakfast. Linda has porridge, milk and vitamin drops. Tracy has scrambled eggs on toast and I have two boiled eggs.

I pack all the waste paper into a cardboard box and venture outside to deposit it in the incinerator right down the bottom of the garden. Weather-wise it is one of the nicest Christmas days I can recall. The Sun hovers above the horizon over distant hedgerows. The white fields glisten and gleam and the air is sharp and clear. A truly white Christmas, just like on the cards and in the words of the song.

Back inside, Tracy prepares the Christmas dinner while I play with Linda's new toys. The shape sorter has made no impression upon Linda at all. I make an effort to arouse her interest by giving a short demonstration. She takes a piano disc out of her mouth and toddles over.

"Oooh!" she exclaims, pointing to the wooden box. I give another demonstration and she sits down beside me to try for herself. There are four cut-outs in the red hinged lid which correspond to round, square, triangular and rectangular shapes. The first discovery she makes is the spherical wooden ball fits neatly into her mouth. Quickly, I fish it out before it lodges behind her teeth – phew! I explain to her the error of her ways, then drop the sphere into the round hole. I try the other shapes, but she is so pleased with the sphere she ignores the rest. The sphere has a soft form and is nicer to hold than the other shapes, which are angular and harsher to hold. Once again I try the rectangle and the cube, but no; if they do not fit through the round hole she lifts the lid and plops them in that way. Clever?

Dinner is ready. The aroma has been making my mouth water for the last hour. We march to the dining room, which is just down the inner hall, and Linda is fastened into her chair. A bib is draped around her neck to keep her dress clean. Tracy brings in the first course – soup of the day – but wait a minute: we have forgotten to pull the Christmas crackers. CRACK ... SNAP ... Plomp. (The last one didn't work.)

"Here's your hat, Linda." Tracy places the yellow crown on her head. Linda looks up to see what it is and the hat slide down over her eyes, blindfolding her.

"Peep-o!" I call, habitually.

Tracy and I don our hats. My Christmas cracker motto reads, "The truth is rarely pure and never simple. O. Wilde."

Tracy returns to the kitchen, taking the empty soup bowls. Linda plays with her hat. I sit bathed in the streaming sunlight, looking through the window at an angle across next door's garden, across fields which gleam virgin white. What a lovely setting this is. It is the kind of scene which etches into the mind and leaves a permanent happy memory imprint for recall in more unsettled times.

The main course arrives and then the pudding follows. We take turns to feed Linda, whose dinner is the same as ours with ingredients chopped up finer, especially the meat.

After gorging ourselves, Tracy and I need time to recover, and there is a mountain of washing-up to tackle. Half an hour later all the chores are done. Now we prepare all Linda's travelling apparatus: high chair, bib and cup, nappies, coat and a selection of toys, to visit our parents.

First stop is my Mum and Dad's house. Linda gives a brief demonstration of her new toys then applies herself to the task of grabbing the chrome and wooden biscuit barrel. We are invited to tea tomorrow, Boxing Day, and naturally accept.

On to Tracy's parents next, where we are dining this afternoon. Ah! This is the life ... temporarily.

Tracy and Linda are welcomed by Dad and Mum-in-law while I unload the back-up equipment. Once again Linda entertains. I observe from a discreet distance, a glass of sherry in one hand and a mince pie in the other. I feel quite content.

Boxing Day is a virtual re-run of Christmas day; a fraction less exciting perhaps, but just as pleasant.

The following day the bathroom sink blocks. This is a positive sign the festive season is about to end. The sink remains partially blocked until the hardware shop re-opens on the fifth day of Christmas when I buy and fit a new U-bend beneath the bowl.

On the tenth day of Christmas, while Linda has a mid-afternoon nap, I take down the Christmas tree. The lights, the baubles, tinsel, chocolates, have all been a great temptation for her. Twice now she has fallen head-first over the arm of the couch in an attempt to touch this aesthetic marvel. I am back in work tomorrow, so the tree is coming down two days earlier than tradition allows in case Linda achieves her target while Tracy is busy elsewhere in the house.

There are lots of pine needles all over the stereo, the coffee table and the surrounding floor area. I grasp the prickly stem and a shower of needles cascades onto the floor and down the front of my pullover.

The poor tree is dying. I carry it outside and stand it at the bottom of the garden, leaving a trail of pine needles in my wake. Its last few days will be spent in a more natural environment. Each pine needle will fall until a bald skeleton remains – as stark as a wooden cross. The tree is being deprived of nourishment from the roots in the most catastrophic way imaginable. Water, full of mineral nutrients, is essential to the healthy tree. It is drawn in at the roots by the power of evaporation created up above at the leaf. On its journey through the tree it nourishes all the cells and tissues it encounters. This is its primary purpose, but some water remains to manufacture sugar in the leaves. Any excess water evaporates out of the leaves and back into the atmosphere.

In the heart of the tree trunk the water column carries all the chemical messages throughout the organism. Leaf, twig, branch, trunk, are all interconnected by this liquid telephone exchange inside the tree. A large fluid brain, if you like, responding with great skill to an ever-changing environment. The tree senses the seasons and adjusts its growth to suit. If a branch breaks, new secondary growth is stimulated below the damaged area in a different direction. When it is dry the leaves regulate the evaporation of water. Balances and adjustments are constantly being made; endless internal activity inside a deceptively static exterior.

Good communication is essential for healthy tree structures. Pyramid hierarchies also aim at good communications between the groups of people residing under the main headings of management and labour. Labour, the industrial working class, can be subdivided again: skilled manual worker, process worker and unskilled manual worker. Each grouping has its own outlook on the world, its own characteristics created by the set of rigid rules which defines the groups working domain. This is where the pyramid starts to fall apart, at the seams, along the boundaries between the sections, where communications are tenuously linked together. The information network, in the pyramid structure, can be envisaged as a thin line of mortar filling the cavities between semi-autonomous sections. When a weakness occurs, more rigid rules are concocted to plug the gap; more cement to bind the monolith together. But this corrective action is only a short-term solution, like a man who borrows money to pay back an existing debt. Eventually something, or someone, has got to give.

It would be nice, somewhere along the line, to rethink the attitudes and actions which we ourselves have produced and maintain.

Creative thinking is a talent the majority of human beings do not utilise to the full, if at all. Humans can be quite lazy creatures. They soon learn certain actions will, in general, produce certain effects. Just as Linda points to the button I have taught her is red when prompted. She does not really understand, but can produce a result, even though it would be beyond her to select another unfamiliar red object out of another picture. It seems once we have compiled a large memory bank of associations we can then omit the tiresome business of creative thought and settle down to a lifetime of automatic responses. Doing what is "right", not what is good. We placidly persist in being consistent even when consistently wrong, because our thoughtless actions are, in some vague way, supported by an arbitrary rule. This gives a feeling of stability. Yet this unthinking attitude slowly leaches away at the mortar line and ultimately causes instability. Inflexibility, the lack of give and take, eradicates respect. When this is lost people polarise to either extreme and vast quantities of constructive energy are lost in destructive conflict.

So good communications are highly desirable and sought after in all good structures. They are also vital when solving tricky problems of expansion: when to expand, in what direction and by how much. Accurate information from every part of the structure must be impartially assessed. In a factory situation this assumes the collection of unbiased facts being made available by each section concerned. Only then can a fitting decision be reached, preferably by a management team devoid of preconceived dogma.

On a more personal level, decisions affecting your own tree of life can only best be made when all aspects are considered. This may seem an obvious statement, but how often do we act on impulse and, for example, buy an item that our financial branch cannot support? Later we have to cut back on something else to compensate. There are no easy answers to problem solving (remember the motto in my Christmas cracker?), but amidst all the apparent confusion there are ways of applying our minds which can clarify problems. This introspection involves visualising, in our mind's eye, a framework for each problem to fit on to, to focus our oscillatory thoughts. Then, now knowing what a good framework looks like, we can invent ways of achieving that end.

When all the data is collated, examine it from all angles, not just from a conventional viewpoint. Try to step outside the framework, as if walking around a tree, and visualise the different interceptions of the branches of thought as the perspective changes. Inspect the problem with an almost alien eye, questioning even the most apparent fact. Only good ideas can withstand close scrutiny, those remaining, even if supported by reams of rules, can be classed as bad and rejected if they contain any logical inconsistencies.

To achieve good communications a free, fluid flow of ideas and information is desired. In practice this is hard to achieve. I think it would help if more emphasis was directed towards teaching the young the part they play and will be required to fulfil in society rather than cramming them full of facts. The youth should appreciate they are part of a vast social organism named mankind and be subtly led away from a self-centred, I'm-all-right-Jack, attitude. Older minds, too, must either be re-directed or abandoned. All must appreciate their place in the order of things. The species must survive the modern threats of anarchy, irrationalism and extremism. Older minds should direct their thinking away from rigid application of detailed rules and focus their attention back towards the overall good. There is so much red tape and regulations about good ideas are suffocated prematurely and novel ideas take so long to pass through the process they become obsolete on the way.

Once upon a time I worked in a department where good communication truly existed. Sounds like a fairy tale? More of that later; but for now let me say the vital ingredient that achieved this happy state of affairs is nowhere to be found in a rule book. It was not taught to me during my apprenticeship, or in college. All rule books and the rest teach are objects to interrelate amongst yourselves as you wish. Sometimes they throw in the occasional convention for relating ideas, but generally it is one isolated formula after another. All the equations in the world, when laid end to end, do not produce good communications. We are taught a list of bland facts quarantined from the rest of mankind's cultural experience, statistical truths without any soul.

The desirable ingredient has already been mentioned: respect. Where does it come from? How is it created?

Respect is formed by an awareness of the other person's problems, their point of view and how eagerly they attempt to achieve results. Such awareness is hard to foster within a system built with isolated compartments.

My father, who spent twenty years as a factory liaison officer, told me how he was expected to identify personnel from different disciplines and get them to interface. People in different parts of his aircraft company like Production Support, Stress Office (airframe stress that is) and Design Office had worked within a hundred yards of each other for years and had never met.

Often there is a feeling we are the only ones doing anything constructive while the rest of our colleagues just mess about all day long. This is hardly ever the case. We think they concoct elaborate alibis for doing nothing, excuses which they adhere to and preach rigorously until we come to believe them. And all this distrust and disrespect stems from the lack of insight into their activities. We never glimpse their vantage point; to do so is anti-pyramid. We must keep, or be kept, in our places, for occasionally some of the hierarchy do achieve little at great expense. So we feel held in an economic trap and produce adequate work for adequate financial reward. Rarely is there any sense of achievement. Job satisfaction nil: involvement zero: tragic waste.

The Christmas/New Year holiday is over and the life of leisure ends. Back to the treadmill, working to pay bills which keep going up because the government tells us we ask for higher wages to pay bills. But there is worse to come.

On return to work, on a typically cold January morning, on the normally cold bus which arrives late, we find, on the noticeboard in work, a letter from the works' general manager announcing five hundred redundancies.

The factory pyramid is under attack, collapsing in the midst of a recession.

Looking at it another way, we are about to undergo a period of pruning.

It was around this time I became aware of the Pareto principle, also known as the 80-20 law. In 1906 Pareto observed that eighty per cent of the land in Italy was owned by twenty per cent of the people. Other facets of human life also seemed to have this 80:20 split. In business, eighty per cent of the sales come from twenty per cent of the clients. The richest twenty per cent of the population receive eighty per cent of the available income. Eighty per cent of crime is committed by twenty per cent of the criminals.

But my adaption of this principle is unrecognisable from the above. I found out, by trial and error, if you had a machine rated at, say, one hundred Watts, and you reduced its output by twenty per cent, its reliability increased dramatically by (perhaps) eighty per cent. This simple technique applied to a range of process plant equipment, like simple pumps, improving reliability so much that any slight fall in throughput was more than made up for by increased running time.

I also think that if people working flat out, as is the modern way, had some leeway in their tasks, they would cope better and both the person and the employer would find sustainable benefit in the long run.

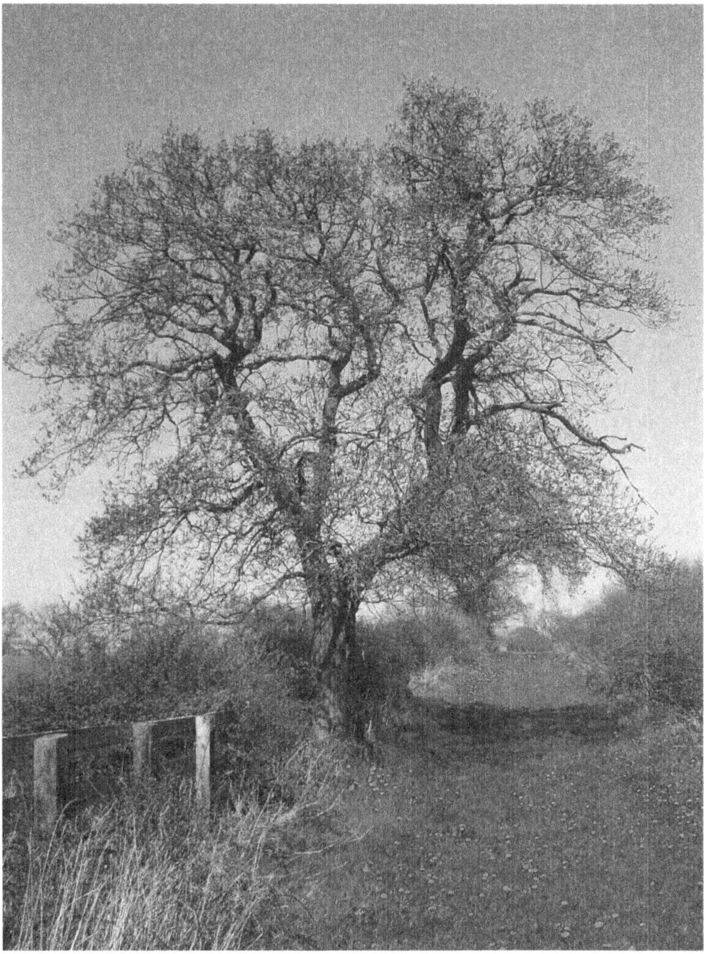

Branch 11

Stock Taking

In Britain, January, February and March are dreary months. The days are short, the nights long and the weekends either cold or wet. It is a good time of year to huddle around a fire, reflect upon the past and plan for the future.

In the garden a lot of debris is strewn over the lawn and the borders. Leaves and twigs and the withered remnants of last year's flowers mottle the garden with sombre colours. The roses need pruning, too.

Now is the time to plan the vegetable patch arrangement for this year's growing season. When we first moved into the bungalow, the vegetable patch was a waterlogged lawn. I imagine this was sown on top of clay which the bungalow builders excavated from the foundations. The first dig unearthed bricks, rubble, roof tiles and the occasional piece of wood. During that first year I planted potatoes because of their reputed ability to break up freshly dug soil. Frankly, I think this is a bit of a myth. The only thing that breaks up the soil is the digging when you plant the potatoes and the subsequent digging when the crop is harvested. We had a lot of funny-shaped potatoes that year: potatoes that had grown around bricks, or flattened ones that had grown against roofing tile rubble.

The second year was much better. Most of the larger stones had now been extracted and cemented into a new path and the winter exposure had made the clay more workable. During the winter I added ash from the solid fuel fire and sprinkled lime onto the land to chemically loosen the acidic clay. The transformation was quite remarkable after this treatment, so in the spring I diversified and planted carrots, broad beans and beetroot, as well as some potatoes again. When the soil was a little warmer I added lettuce and sprouts. Only the lettuce failed, although the sprouts did not amount to much.

Last autumn the contents of the compost heap were buried in the plot. Since then it has been dug again and dappled with more lime.

You need patience when gardening. It can take years to create an environment suitable for growing the best crops. Gardening cultivates patience as well as crops. The solution to some problems can take years to filter through a system before any tangible result can be seen; the bigger the system, the slower the process.

Fittingly for this time of year, I am now going to patiently take stock of what I have unearthed so far during my leisurely reflections on some aspects of systems. This is necessary so we can now focus on future applications of tree structures, as opposed to pyramid hierarchies.

I began last autumn when I tried to illustrate the subtlety of the problem before us. I embarked upon a typical day and discovered a lack of involvement pervading our everyday order, typified by the neglect of public artefacts. As if they had no owners, whereas they are owned by all of us. No one seemed to care any more.

I observed rigid geometry was rarely found in nature on a macroscopic level and did not occur naturally, but was a set of man-made conventions, useful, yet artificial.

I introduced you to Linda and found she exhibited no inherent preference for geometric shapes. She would have to be taught to like them; she would be surrounded by them (in any case) and consequently their style would soon become more familiar to her than nature's backcloth.

Then we entered a factory to investigate their administrative structures. These turned out to be based on a pyramid hierarchy, geometric by design, and I gave an example of this in the form of the organisational framework of a workshop. I discussed the four levels of thought needed to maintain pyramid hierarchies and how misunderstandings lead to sheer bloody mindedness.

I wondered why organic man used rigid geometric structures of organisation to control his activities, sensing this was the crux of the problem. The reason he no longer cared was a consequence of being subservient to an impersonal system.

This led me to look for a structure more suitable than the pyramid for man to base his activities upon. Preferably a structure which:

(a) Permitted more flexibility, yet had a specific shape when healthy.
(b) Had a definable beginning – a heritage.
(c) Provided a goal to aim at, and
(d) Could be explained more easily than pyramid structures, by use of analogy, to both old and young.

I described my feelings of depression at work when undergoing a bout of the going-nowhere blues, then I applied myself to analysis of this problem. The result was a graphical representation of my career, which I termed a Career Fork. I soon realised this would not be relevant if considered in isolation, so added a Family Fork alongside. In the end I constructed a structure the shape of a tree, each branch representing a path I had arbitrarily chosen to take. More questions emerged: what shape was a good tree? What is its goal and its origins? I turned to nature to look for answers.

This was a satisfying period of my investigation when observations of real trees helped to resolve the above questions. The shape of a good tree was spherical with no unstable extremities amongst the branches. The goal of the tree was even, all-round growth. The tree was rooted by its past; a good heritage was an important factor in a healthy tree.

I then compared the communication medium in pyramid structures and in trees. The former comprised of messages making quantum leaps from one discrete level to another –

digital. While in the latter the flow of information was fluid – analogue.

And that brings us up to date.

Each chapter can be considered a branch of thought covering a certain area of the theory, a compilation of ideas and observations, page upon page, like leaves on a branch, separated in space, yet all closely connected to the main trunk of thought.

So far we have seen some of the non-productive areas rigid hierarchies generate and (when encountered) how frustration sets in. Yet I have found from personal experience to understand even a little about pyramid hierarchies can help enormously. There are numerous examples I could use to illustrate my meaning. The first that comes to mind relates to our bus failing to pick up passengers. Remember, our "factory" bus is a private bus and it does not pick up old ladies, ex-servicemen, or anyone else, even if it is nearly empty. When observing this scene you can feel the animosity focussed on the driver as the public relations catastrophe runs its course.

Another area where frustration is encountered is in the factory. You may be given a job which you feel is over-designed – unnecessarily complicated as the purely theoretical designer demonstrates his flair to his admiring colleagues. Building something you feel is going to be inhibited (overridden) as soon as the operation personnel see it is a form of slow torture.

Conversely, the designer no doubt sees some of the craftsmen in an unfavourable light. He probably complains they cannot build his ideas since they appear to be simple and have too many tea breaks and are always querying (yes, impudently questioning) his simplest design, almost as if they know something about the job. I think he sometimes gets angry, too, but until he writes his book we will never really know. That is part of the trouble; we do not fully know how others feel.

These two examples have one main thing in common. In both cases the individuals involved become a focal point for personal abuse and enmity, yet they find themselves in situations not of their own making and are devoid of most of the blame.

The bus driver is instructed not to pick up the general public on a private bus because it could violate an insurance claim in the event of an accident. He is bound by a rule laid down not by anyone who wishes to promote bus travel but by someone who deals in insurance, whose main desire is to buy a new Volvo each year so he doesn't have to travel on those damned buses with the impolite drivers.

There is no point staring daggers at the bus driver as that is bad for one's blood pressure. When you realise the driver's actions are truly nothing personal and he is only the unfortunate entity who applies these rigid rules in the face of ex-servicemen who fought in a war to keep the country free from fascism, then the feeling of ... well ... sympathy begins to gel.

You have to be emotionally hard to tell old ladies they cannot travel on a half-empty bus. I mean, it defies logic and bus drivers have all had mothers.

Most designers have fathers as well as mothers. Like the bus driver, the designer is obliged to comply with numerous regulations not of his own making. He has to design rigs that are so safe, and hence complex, they sometimes become monsters. Consequently the installation team become dismayed and frustrated, focussing their hatred on the designer himself. On completion of the rig, the process operators take over. Soon it may become apparent the rig is so safe it is usually in alarm condition. This is undesirable. Remedy: eliminate all overdesigned features and increase alarm actions to a sensible level. The responsibility for sanctioning these adjustments falls in the process managements court. Now if anything goes wrong they are to blame, not the designer.

Usually things go better. The tradesmen, who spent hours trying to fathom out the intricate design, are subsequently sent to eliminate it, in effect producing a more practical rig. Meanwhile the designer continues through life thinking he is good. Does anyone tell him otherwise?

The above situation is a result of the poor communications inherent in pyramid structures. The rig is conceived and designed at a level where the people who have to build and maintain it – the practical people – are not involved. In fact, their experience is totally ignored. So when design errors occur (a trivial slip of the pencil perhaps) a minor fault is transformed into a major cause of resentment. The installation team, who had no part in the discussions, feel no great inclination to correct the error. Instead they calmly send their adverse findings back along the Chain of Command and phlegmatically await an obvious reply.

The pyramid structure is impersonal. It alienates people from people by putting up imaginary barriers between them. Instead of the emphasis being on completion of the rig as a whole (the goal), attention is focussed on the integrity of each little working unit. The project takes second place.

We seem to have adopted a cellular, insulated life style in the West, in keeping with our blind subservience to pyramid systems. Each day is fashioned piecemeal, a jumble of random elements to which we automatically respond. Our lives drift along a zigzag course, blown by the winds of change and limited by the passage of time. The big objectives lie beyond our everyday experience, so we naturally never give them a second thought.

Right now I feel these attitudes are a fundamental cause of much unrest. They pervade the whole of society and need to be altered, but how?

Some people think change requires violent revolution, a militant show of the peoples disgust towards the Establishment. It is rare (in the West) for things to get so bad as to unleash this solution. Under more normal circumstances revolution is an extreme answer which cannot be easily condoned. Besides, revolution is not progressive, it is regressive. A new regime still has to create order, build a system, and it starts at the top and builds down. Unwittingly, the apex to base, unidirectional, geometric, rigid, top-down system is reborn under another name: new name, same old system beneath. Another pyramid emerges with similar characteristics as the old one. Forgive me if I sound cynical here.

What I want to see is a little bit of evolution. Take the existing pyramid and alter its structure to create a superior model. The new model will need new elements to fashion a more holistic and continuous framework by people who can see and want to see beyond their boxes; above, to the left and right and below as well. People whose personalities I would like you to visualise as expanding spheres, not coffin-shaped boxes. Let's start with the man in the street.

If we talk to the bus driver, or the designer, outside of their normal working environments, we may find they are not ogres at all. Once removed from the confines of their career boxes they become perfectly normal Earthlings, concerned about the same matters as you or I and are either baffled or divided about the solution to these problems. Like the rest of us, they may have difficulty paying their bills, think they pay too much tax, complain about the lack of discipline in kids and the amount of litter on the streets. They are of the same species, mature beings, fairly set in their ways. It is not, I trust, beyond them to alter their ways and attitudes a little bit: to rekindle inside them a feeling of friendship and sociability, reawaken their interest in life and arouse their drowsy curiosity, all feelings they once had when they were young, before they were taught the techniques of Western civilisation.

Each of us can begin rearranging the pyramid right now to make our working lives more acceptable and our family lives more complete, in harmony with man's natural organic inheritance. A heritage, a culture, based upon open programme learning; a fixed instinctive response to certain situations combined with memory space to be filled by learning, the latter providing the flexibility to adapt to a changing world. We must utilise these gifts to the full, even if this all sounds too utopian.

The changes required to convert pyramid hierarchies into tree-structured frameworks are not all that drastic and I believe the goodwill necessary to achieve a transformation is still there, lying quietly dormant deep within each of us. It was embedded there eons ago at a genetic level.

For millions of years man led a hunter-gatherer existence where co-operation, not conflict, was the key to success. Man and nature were in equilibrium, and then along came agriculture to upset the balance. This brought rapid population expansion and with it material possessions which needed protection. So only eight thousand years ago the world's first armies were assembled. In terms of man's evolution this is only a very brief period of time. Our genes contain more remnants of co-operation than conflict.

With these thoughts in mind, I will tackle the first system transformation.

Earlier, in *Branch 4*, I showed you a pyramid with four levels (as below) which could underlie the production of almost any product. In our case the end result was a test station, also called a rig.

| Customer |
| Design Team |
| Management |
| Skilled Manual Worker |

To recap, the customer describes his needs to the designers who act as an information exchange. They organise the schedule and materials for the job based upon their ideas.

The plans are fed to the management for their perusal and finally the tradesmen are delegated to build a specific part of the rig. But somewhere between the tradesmen and the designers is an area devoid of effective communication, a layer of mortar within the pyramid so weak and tenuous it can accurately be described as a vacuum. If I had to invent terminology to label this gap then "the staff/industrial interface gulf" might suffice. On one side of the chasm are the theoreticians; on the other side the practical men. The latter feel little sense of association with the job. They are rarely told specifically what it is for, how it should operate, or its purpose. It is said they have no use for such information; there is no need for them to know. It is above their heads and anyway, the design team can look after all that. But that is pure theory once again. When people have no responsibility they become irresponsible. Likewise, if there is no demonstrable interest from above they drift towards indifference and the quality of their endeavours plummets. The result is an inferior job. No matter how good the design is, good practical work is still highly desirable. Regrettably, this is not reflected in either the encouragement given to the skilled workforce, or their wages relative to the designers. Consequently companies get what they deserve: overdesigned and moderately made assemblies.

The designer does have certain financial constraints to limit his flair. Usually the only feedback of a practical nature occurs after the design is finalised and the job built; when it is commissioned – too late for major alterations. The practical knowledge is added after the job is built, not as one might logically expect before the hardware is assembled.

These are some of the inadequacies I feel are evident when constructing a rig under pyramid hierarchies. No matter how good the people involved, well-educated charming people, as soon as they are placed within the existing framework, it (not they) creates friction. This system truly is to blame for these faults and tensions.

The transposition from pyramid hierarchy to tree-structured frameworks is achieved in this manner: construct the existing pyramid hierarchy and invert the model. Give it continuity and eliminate the boxes. This is harder to illustrate but will look similar to the sketch below:

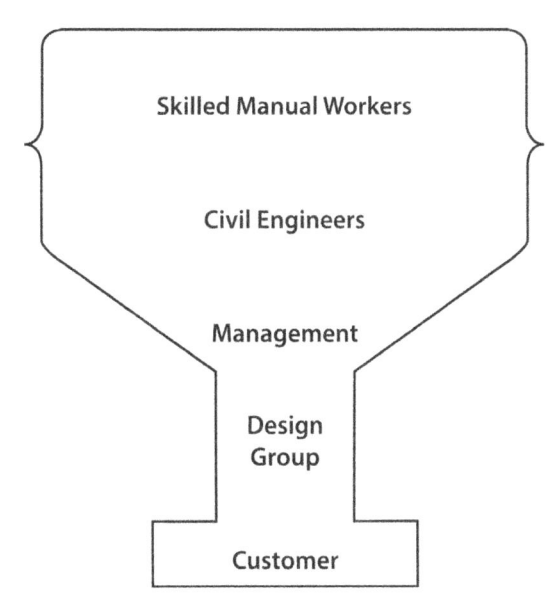

```
Skilled Manual Workers

Civil Engineers

Management

Design Group

Customer
```

This new two-dimensional model tree structure will have to suffice for the purposes of this work. Yet remember a real tree structure has three dimensions, it has depth, too. Elements which are drawn on either side of the two-dimensional tree could be closely related in a three-dimensional model, just like a European map of the world shows America and Russia at opposite edges far apart, whereas they meet in close proximity across the Bering Straits. Three dimensions need three projections, which are too cumbersome for our purposes.

We are therefore left with our two-dimensional drawing of a tree structure to examine for excesses and deficiencies.

In this analogy the customer is the tree's roots. The ideas in his mind are shielded from view, almost as if hidden below ground. He provides the resources to nourish the tree.

The tree's trunk is the design group; they should be the main support through which all the information flows towards the crown. Top management assess the tasks and delegate them to a particular trade. The branches subdivide again. Middle management now set the individual tasks to the engineers and tradesmen themselves. The latter are at the twig end, the outermost reaches of the structure.

The customer provides a tight specification for the designers. They have a few options, differing ways of achieving the objectives. The management have to choose the best way to arrange the tasks from several strategies. Whilst at the twig end, each trade can freely transgress towards areas occupied by other disciplines. The farther up the tree, the more flexibility, this is the inverse of the current pyramid approach.

The demarcation disputes pyramid hierarchies create meld into an overdue oblivion. Twigs are allowed to partially transgress into other craft areas, intertwining, whilst still attached to their own particular branch, cross-pollination (if you like) giving birth to fruitful interchange.

Is this flexibility all a tree structure has to offer? No, not at all.

Fundamental pyramid thinking states, quite clearly, all information is exchanged in the upper layers then fed down to the base in a unidirectional process. I quote from a contemporary magazine stressing the importance of good communication: "It is top management's job to feed information down the line and make sure it does not become distorted." Note the implied one-way direction of idea exchange and also the desire not to have the information altered. The latter, of course, is only possible if feedback is eliminated.

In tree structures "top management" resides nearer the base of the tree. This may seem like a lowering of status, but in fact the reverse is the case. To function near the bottom of a tree's structure is a measure of maturity, experience and esteem. In pyramid hierarchies, those at the top can and often do function without feeling or regard for those below. In tree structures they are required to continually nourish those above and to take notice of the reactions of the humblest twig.

There is one facet of trees I have not delved into so far: the leaves. We have only seen them strewn on the autumn ground, crisp and desiccated, awaiting decomposition.

When I began this saga, last August, autumn was rapidly upon us and I never found time to describe their function. Now that spring has returned, the new leaves are bursting out of their winter wombs. Soon they will feel the prevailing winds.

Leaves are the tree's lungs, breathing the air and converting it to food for the living tissue. In the leaf's cells the water from the roots combines with the carbon dioxide in the air and the radiant energy of the Sun in the process known as photosynthesis. The tree's main source of food, carbon dioxide, is gathered at the leaf end and is not derived (as one might think) in the roots. So the tree is fed from both ends of its structure, yet its *primary* food source is gathered at the twig end.

To apply this kind of bi-directional system to rig design would be, to say the least, innovative by today's standards. Yet this has not always been the case; for in days gone by the craftsman was the designer as well as the doer, and this unity of thought and skill created artefacts of great quality seldom matched by *any* skilled team of today. Suggest to a present day designer that the craftsman should have a say in the design would more than likely result in a fit of apoplexy. It is just not the done thing and has not been since the day man started to specialise to the extreme. To divorce theory from practice is a logically absurd way to build anything. It can only be achieved with persistence. Then you end up with careless craftsmen and disdainful designers. This man-made dichotomy needs to be resolved before the decaying pyramid hierarchy of the factory can be converted into a healthy tree structure. The interface between theory and practice must be blended together until it melts into a smooth transition: a zone in a branch where there is a free exchange of information in both directions; in pyramid terms, not just down-the-line, but *up* it, too.

Remember, in my analogy craftsmen reside at the twig end of the tree structure, up where the canopy is buffeted by the wind of change. At the top they feel the climate and are very vulnerable in adverse conditions. Taking the first aspect, the craftsmen can feel, sometimes in a physical sense, how the job is going. On the second point, he is amongst the first to be affected if the tree is in any way damaged, even though he has no say in which direction the tree's structure is being driven.

This "wind of change" I keep throwing at you needs further elaboration. I would like you to imagine it as a flux of ideas and opinions in which an individual's tree of life, or the factory's structure, are immersed. To a large extent these can be collectively conceived as political streams of thoughts, reflections of ideologies as yet unproved; or sometimes the present day collective unconsciousness of mankind. The wind is rarely still. It swirls around gently rustling our daily lives and the poor are rustled more than the rich. The latter can buy and live in their own cocoons of climate and can indeed affect the wind direction themselves. Sometimes two winds, born in different pressure areas, collide. Then turbulence flexes our lives and we have to redirect our activities to bear the brunt of the storm.

Each day the humble electrician enters the wind; sensing, observing, building and modifying a mental model of the surroundings in order to formulate a response to them. Man is not indifferent by nature. He needs an outlet to channel his reactions; to relieve him of the unsolicited knowledge that accumulates. He wants (oh, how he wants) to give something back to the system. He wants to improve it. He wants to nurture it. But mainly he just wants to feel a part of it.

He wants the little box he inhabits to have arrows coming out of it joining up to other boxes with arrows coming out of them. Instead, here at the pyramid base, just one arrow plunges into his cell – straight down-the-line. Like an arrow to the heart, killing his interest, puncturing his hopes.

Thus my diagnosis of pyramid hierarchies is one of terminal decline. The cure is better communication. But how can this be achieved?

As I review this branch thirty Earth orbits later, I can sense the working world I described may well be a thing of the past. I worry perhaps I have protested too much and my overall case will be diminished. I ask you stay with me a little longer, as this was merely the means I have tried to use to demonstrate a conversion from a rigid system to one more fluid; a practical example, as it were.

Remember, these were the days of big factories with thousands of employees. Individual departments could have one hundred people on a single project. As the years rolled around projects became fewer and the size of the teams became smaller. Yet still the project teams may have been the size of a single industrial unit on a modern enterprise zone. I wonder how these organisations are managed. Would they see anything familiar in this last branch of thought, or are things running smoothly now? Do they have free two-way communication? Does everyone feel a part of the project?

Also, my panel of manuscript reviewers suggest I should use non-gender specific descriptions rather than "tradesmen" and "craftsman". In my defence, the trades were totally male dominated when I was writing the original manuscript and if I had used "tradespeople" contemporary readers would have raised an eyebrow.
Meanwhile back in the factory ...

Better communication is the order of the day and, obviously, in order to communicate people have to get together and talk, frankly and freely. So meetings are arranged and people appointed to pronounce on the issues concerned. Such a gathering is called a committee, a title which has gathered a lot of distrust by the population at large. You have probably heard of the description of a camel as being "designed by a committee". It seems an unlikely animal, even though it is perfectly evolved in its desert home. There may be little faith in committee-style decision-making, yet there is no better way of communicating than by bringing together all the people involved.

The British political system has, near its current apex, over nine hundred committees to co-ordinate all aspects of government control. Each committee can be ultimately traced back to the government's main committee, called the Cabinet.

The trouble with committees is not the concept they represent, it is the people who sit on them. Committees are magnets for the word merchants, people who do not actually achieve much, but (by heck) they can't half talk a good job. Meanwhile the chap who really is good at the job is so busy covering for the rest (who are at meetings), he cannot be spared time off to attend.

I am not sure if committees will improve much in tree-structured organisations, but what I am proposing is one extra committee – the canopy committee. What is its purpose and how will it operate? This is the idea ...

In pyramid systems the elements making up the bottom layer have inadequate outlets. All the units above have information passing through, in and out (yet still, more often than not, one way) which can be modified by the element concerned. At the pyramid base messages are delivered, then the messenger goes away. Sometimes he is advised to go away. Sometimes he says, "Right, I'll leave you to it."

Although it is relatively easy to find the messenger again, understanding the message he has delivered requires an insight into the thoughts behind the ideas. These emanate from higher levels. Consequently the response at the base to some messages may be inappropriate; out of context. Likewise, the commands from on high may be incompatible with the actual physical reality down below. It is a classic pyramid problem: each unit has difficulty seeing beyond its own perimeter. The overall objective is obscured or vague.

We left our tree structure looking like no more than an inverted pyramid with its only asset a greater degree of flexibility. That in itself would be a considerable improvement. Now we seek to improve the communications by providing feedback from the tree's canopy down through the main branches to the trunk.

What follows is the basis of a canopy committee within a factory format. The concept though could be applied to *any* tree-like, bottom-up, two-way structure of organisation.

All committees need a chairman to control the meetings. This person must be selected from outside of the purely practical or theoretical domain so he may impartially decide upon the polarized opinions with which he will be confronted. He should have an overriding interest in progress on the project under discussion and a desire to smooth out the difficulties that inevitably arise. Hence my candidate for chairman is drawn from the production side; preferably a senior production manager, keen to see the project in his care operating smoothly and with a healthy disrespect for so-called designers and craftsmen.

The chairman's secretary, who makes a record of each meeting (the "minutes" as they are known) should also be production oriented, though it is not necessary to involve a senior person here. The committee's members are now drawn from the outermost twiglets of the tree structure – the canopy of the tree. Two representatives of each trade attend: two fitters, two electricians, two instrument mechanics and two production workers, for example. No self-pronounced experts are permitted unless the committee invites a knowledgeable person to attend to clarify an aspect of the operation. So the normal attendance comprises all of those who are not directly represented in the old pyramid hierarchy. They can speak freely on equal terms, without fear of reprisal, directly to the man who will be responsible for operating the project. Any ideas or suggestions which are conceived at these discussions are still attributed to the originator and can be pursued through any suggestion scheme the factory may run.

The committee's activities would not replace normal managerial responsibilities. Management still have the final say in all actions which the committee decide upon. There is no diminution of their authority or status.

A typical meeting will proceed like this: the chairman exchanges a few pleasantries with those gathered and then hands around copies of the agenda. This follows a standard pattern and is similar for each meeting:

(a) Apologies for absence (there is usually someone who cannot attend).

(b) A re-run of the last meeting's minutes, recalling the previous discussions.

(c) The chairman now informs the meeting of any news affecting the production and discusses the progress of various rigs.

(d) Time is set aside for other items the members now wish to raise.

The meeting then ends.

The canopy committee members do not discuss the high-level decisions about rig finance and conception. They discuss relatively minor flaws which have slipped the attention of others further down the tree, but which are, nevertheless, detrimental to healthy operations. Occasionally an idea is unearthed which can save a considerable amount of time or materials. Then all feel the committee is more than justified. But normally the matters raised are of lesser priority, though this is less important because while the electricians and fitters and the rest are sitting there something else, almost magical, is happening. An invisible alchemy is beginning to erode the edges of the individual boxes and each can see farther into their fellow craftsmen's trades. Unwittingly, they are transformed from a gathering of solitary specialists into a team with mutual respect for each other. At the same time a feeling of involvement is being generated. No longer do they feel like numbers on a payroll: this is their rig; they built it and by Jove, they will make it work! They will show these designers what true ability is!

What of the designers? Why are they not invited to the canopy committee meetings? After all, these gatherings were supposed to help eradicate the staff–industrial interface gulf, yet no staff (except the secretary and chairman) are present. Should not the designers be there to defend themselves if necessary?

No. These meetings are not set up to attack anyone, so no defence is necessary. Their aim is not to drag anyone down; rather the aim is to elevate the tradesmen's interest and involvement.

Most electricians don't want to be designers. If one does, he goes to night school, gets a certificate and applies for the next vacancy. We must accept that there will always be those that want to design and those that want to build. The staff–industrial interface gulf is not created by this fact; the gap is fashioned by the upper layers (the staff, collectively) ignoring the experience of those on the pyramid base (the industrials). As with other cases I have mentioned, there is no premeditated malice intended by the staff. In fact, most do not realise a problem exists and have given it no thought at all. The gulf is, if anything, a traditional, custom and practice. The tradesman's response to this division widens the gap farther. He becomes frustrated, angry at times, uncommunicative, then sheer bloody-minded. Feeling left out, he becomes indifferent, but being left out on purpose turns him into opposition.

Designer and doer are perpetually set at each other's throats and if brought together at a meeting then upset would be the most likely outcome. Again, this is an enormous waste of effort.

The canopy committee tries to redress the balance of involvement, bringing the tradesmen's level of commitment up from nothing to ... well, the sky's the limit. He can become as interested and involved as the designer – easily. No longer an inanimate unit at the pyramid base, now he is a growing part of the tree.

Records of each meeting are distributed to senior managers and a few copies are pinned on relevant noticeboards. Everyone can read what has gone on and, in this way, is kept up to date and perhaps on their toes as well. Who's going to grumble about that?

Fortunately, I can report on similar meetings held upon canopy committee lines of thought. Once upon a time I worked in a department where there was little frustration, no anger and friendly comradeship prevailed all around. Sounds like a fairy tale? The plant ran smoothly and when it broke down we were keen to fix it – all of us. The electrician would isolate and remove cables from a machine, pass the fitter the spanners and help him lift a new unit into place. Reconnection would be made to the plant and power turned on again. Then we would consult and assist process people who tested the new apparatus.

At the time everyone thought the harmony was connected with regular monthly meetings held by the production manager. But nobody really understood how such a simple gathering could produce such a cohesive workforce. I, too, pondered upon this mystery, but only now do I realise where and why these meetings actually fitted into a system. What we had experienced was akin to my concept of a tree structure being fed from both ends. We had become involved in the system which, for a century or more, had alienated and suppressed the craftsmen. Instead of being treated like a necessary evil, we felt wanted; useful. It was a nice feeling.

The kind of things discussed at the meetings could be broken down into three main headings. On a percentage basis the list was as follows:

1. Information and communication: thirty-six per cent.
2. Plant operations: thirty-two per cent.
3. Safety matters: twenty-four per cent.

The residual eight per cent were discussions of working conditions (two per cent) and other miscellaneous items (six per cent).

The chairman stated during one of his yearly reviews "the ideas and observations from the shop floor made the decision-making process more efficient."

He had sampled a broad canvas of opinion (some canopies are made of canvas) and of the one hundred people in the section, seventy-six had attended voluntarily. Several of the committee's members commented they were now more aware than ever of what was happening on the plant.

Strange as it may seem, these meetings tend to be self-destroying. They improve the working atmosphere so well that after a while there is no longer a need to meet formally. Informal chat takes their place and fewer items appear on the agenda.

Running concurrently with the reduced necessity to meet is a period where the plant hits the peak of its reliability. If we draw a standard reliability "bathtub" curve and a curve representing the frequency of meetings, the two would show similar characteristics at first.

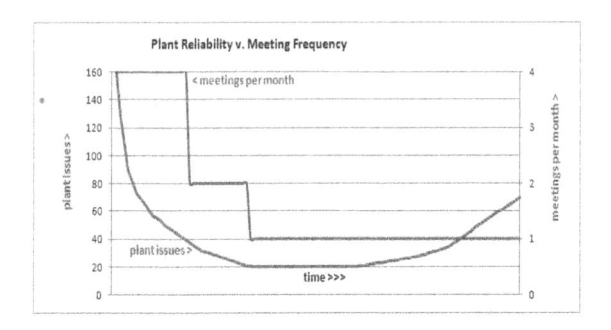

Plant Reliability v. Meeting Frequency

Initially there are always teething problems and plant outage is high. This is termed "infant mortality". Now the frequency of the meetings is high, say one per week (or four per month in the graph). As the plant starts to operate smoothly (when the teething problems are ironed out) there is less demand for meetings and their frequency reduces to, say, one per month. Eventually, after perhaps ten years, the plant starts to wear out and failures increase, but the meetings do not follow suit. Instead (from my experience) their frequency declines until none are held. The project ends. The committee members are dispersed to new departments. Here the need to have a canopy committee may not be prevalent, but their memory is always there lying dormant. Then perhaps one day an individual will be confronted with a situation which forces his subconscious memories out into the open and the idea is reborn. Sometimes a new canopy committee (call it what you wish) emerges and the cycle begins again. On other occasions the idea meets resistance. Committees, factories and trees undergo cycles of growth and consolidation.

Sometimes there is great activity, good prospects and hope. At other times there is less activity, an uncertain future and pessimism. These are natural cycles; rhythmic and unforced peaks and troughs. When one accepts the cyclic nature of activity it becomes easier to understand events at the bottom of a trough. It is far better to accede to the fundamental frequency of these life forces than to try to extrude them from society.

Governments have proposed workers should sit on the boards of big businesses. Occasionally labour officers or managers have a rush of blood and attempt a philanthropic gesture, creating meetings for workers to participate in. But these surges of communication must be forged together at the right time. To try to encourage them in a winter setting, when the system is hibernating, is destined to failure. The timing and frequency must be right. Prior to an increase in activity, hold meetings. When all is running smoothly, have fewer meetings. There is no need to adhere to a fortnightly or monthly routine.

The average company worker does not want to sit on the company board. He just wants to do a reasonably good job for a reasonable wage; but as well as that he needs a sense of involvement. Instead, he is regarded as a number on a payroll, an arbitrary unit of labour. The canopy committee brings the industrial worker within the framework and allows his feelings and experience to be transmitted through the management branch back to the main design trunk at the heart of the organisation. A flow of information and feedback as smooth as is humanly possible may be achieved. The individual benefits and so does the company. And if the individual goes home happy his family benefits and eventually society as a whole is enriched.

The seed of involvement grows until he actually starts to care for things again. He picks up litter from outside his gate; he has the confidence and audacity to talk to his cellular neighbours. All the time the box perimeter is being eroded. He has a feeling akin to intoxication. To use the vernacular he feels "out of his box".

The invisible walls of caution crumble before the enthusiasm of his newly-found freedom. Unleashed is all the suppressed and wasted energy previously bound by the illogical habits of convention?

The bus company could alter their attitude: relax the red tape and the specific rules now rigidly applied. That might make the bus driver happier. He will be allowed to give his mother a lift on a half-empty "private" bus, which will make her feel important, too.

The small gestures gather compound interest. Their beneficial effect can snowball and grow. And all this comes about not by violent revolution, nor by wiping the slate clean, but by merely turning the existing blackboard upside down.

This inversion does not destroy the pyramid. It is far too strong for that and this is not the object in any case. No; our devotion to pyramid hierarchies, whether intentional or not, has made our country one of the most stable in the world. But now they have had their day. New ideas are desperately needed. There needs to be some evolution.

I see these adjustments as fine-tuning a good system to improve the reception for the people. Then the rules may become our servants again and not, as is now often the case, our masters.

The wind of change is a fascinating subject in itself. How do real trees respond to the constant buffeting of wind? Again we need to visit the hedgerows and see what effect the prevailing wind has on their form, though I confess that *more study is needed here.*

On the west coast of Wales the wind blows raw off the Irish Sea inland. It howls up the headlands and into hedgerows and the sparse trees on the skyline. There are no trees on the hilltops, only some stark modern wind turbines, but frequently forestry plantations have been added to the valley slopes. The few exposed trees on the skyline all bend away from the prevailing wind, which is from the north-west. Just like people when exposed to prevailing political dogma, the trees make a concerted effort to shield themselves from the consequences.

The image of the tree on the front cover has the prevailing wind blowing from the left-hand side. The trunk leans slightly to the right and the more sheltered branches aloft, at top right, are also taller.

Yet there are signs that trees also try to fight back. Though leaning away for shelter, they nevertheless thrust out lots of new growth into the oncoming storm; actively trying to oppose the prevailing view thrust upon them from outside.

← This tree has been sheltering its neighbour from the prevailing wind blowing from the left of the page.

Branch 12

Enter Joe Public

April arrives and the climate warms up. We have endured a long cold winter with temperatures below freezing nearly every night. The days were pretty chilly, too.

The solid fuel fire has done a tremendous job. God bless low-technology heating! When coal stocks dwindle we could still burn rolled-up newspapers and magazines and wood. The price of extra coal is prohibitive. It is about time someone mentioned to the Coal Board you can buy two pullovers at the local market for the price of one bag of anthracite.

The days get longer, the evenings lighter and the grass starts to grow again. Yellow forsythia bushes speckle the avenues and daffodils bob about in the invigorating spring breezes. Primroses give an abundant display of delicate flowers and crocuses pop up in lawns and borders, illuminating the cold soil with splashes of gay colours.

During the last three months, when the weather and my stamina have permitted, I have been quietly preparing the garden for the summer onslaught by Linda. My father, who retired recently, has been busy making three gates to improve the garden's safety. One will isolate the greenhouse area and another blocks the path behind it, which is adjacent to the field next door and intervening ditch. The third gate will prevent access to the top half of the drive leading on to the road. All are now fitted.

While my dad has been busy installing the gates, I have moved the water tub off the garden to a safe location behind the garage. The tub is supplied with rainwater via the garage gutter. It was quite a feat of engineering to arrange the pipe to zigzag around the garage corners.

In front of the hedgerow I have erected a plastic coated wire mesh fence. This covers all the little spaces devoid of growth through which an inquisitive little girl might squeeze.

Finally the old rusty fence between our driveway and our neighbours has been replaced with a decorative looping wire mesh stretched between new wooden posts.

So here it is: the safest garden on the estate. Let's follow Linda outside to see what loopholes she can find in the defences.

The weather is cool and dry. Linda is clad in her one-piece playsuit (I've dubbed it her "spacesuit") and a pair of bright red wellies one size bigger than her feet.

I open the back door and lift her over the step onto the cement drive. She giggles with excitement and shuffles off towards the car parked a little lower down the driveway.

"Olooh!" she exclaims, pointing to the car's rear lights which are level with her eye line.

"That's the motor car," I pronounce.

"Bwum-bwum," she replies, then heads down the drive towards the garage door.

The garage door is protected from the snow, frost and rain by a cover made from the bottom of a one-gallon plastic container. It is crude yet highly effective. The cover is hinged at the top. Linda lifts the cover and giggles at her achievement; then she knocks the cover down again. The rusty hinge squeaks and the cover goes "bang" against the handle.

"Bom," says Linda (her version of "bang") and a grin stretches from ear to ear.

She's off again, but now for the tricky bit, down a step and a shallow incline on to the garden path. Oops! She's slipped. More practice needed there.

"Oahh." Linda looks at her blackened, soil-covered hands.

"Never mind," I mumble automatically. No real harm done.

It is fortunate children are designed for falling, close to the ground with a low centre of gravity. Linda is two feet, six inches tall, which I imagine is probably several hundred millimetres at least. Added to that, the padding in her spacesuit makes her fall a safer one. I lift her onto her feet again and brush the dirt off her knees. Off she canters again, her little legs not yet in full control of her wellies.

"Keep on the path," I instruct and she immediately scuttles onto the wet lawn. I forgot she doesn't know the word for path; we have hardly seen it all winter. I lift her back on to the cement and explain that this bit beneath her wellies is the path.

She toddles towards the greenhouse, now protected by the new gate built by my father.

"Oooh." She points through the vertical lathes of wooden gate. She probably remembers a few visits to the greenhouse at the end of last year. Then I used to carry her down the garden to see the tomatoes and sometimes she would grab a leaf while I looked elsewhere.

"Okay." I submit to the barrage of noise. "Let's go and have a look inside." We enter the greenhouse.

All the slug pellets, powders and potions are on the top shelf. Only plant pots and the dry, dusty soil are within playing distance and soon she is having a marvellous time sitting on the metal walkway filling a plant pot with dry soil, then tipping it out again.

While she is playing so contentedly ... ("Linda! Take it out of your mouth!") ... I tidy the work table and arrange seed trays in a row ready for the start of the growing season. I keep one eye on junior's mouth at the same time.

My back garden has a fairly typical layout. It is divided into various zones by the paths; each area has a different requirement and function.

The vegetable patch is a production area and the lawn a place for play or relaxation. The lower half of the garden has poorer soil which I have never found time to attend to properly.

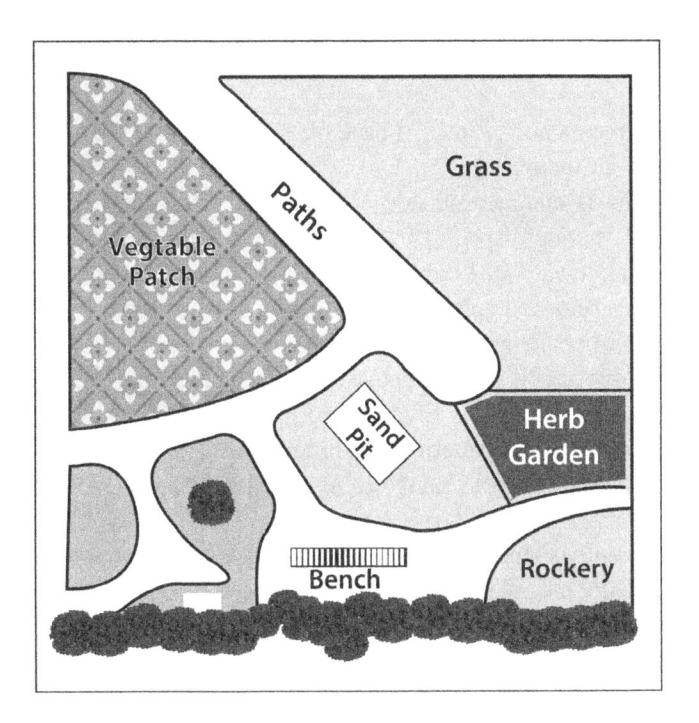

[Garden map with labels: Grass, Paths, Vegtable Patch, Sand Pit, Herb Garden, Bench, Rockery]

The right-hand sector is supposed to be an herb garden, but I have noticed that only the mint and the sage have survived the severity of winter. The left-hand island contains a variety of conifers to camouflage the incinerator behind.

The rockery, in the right-hand corner, covers a big gap low down in the hawthorn hedge. It isn't properly stocked yet, but the potential is there.

Looking at the garden now, after three years' toil, it is hard to remember it was hewn from a common clay pastureland, the remnants of which are only a few feet below the surface.

Meanwhile, back inside the greenhouse it is warm. The Sun raises the internal temperature above seventy degrees Fahrenheit and consequently I begin to sweat. I open the sliding door a few inches to provide some cool ventilation. Linda's chubby cheeks are red; she must be warm, too. Her spacesuit, however, is turning a dull brown colour from messing about with the dry soil.

"Come on now, Linda." I have made a decision. "We've done enough for today."

Linda is marched outside on to the path and the feeling of warmth is soon dispersed by the cold East wind. We make our way back towards the kitchen door. Linda dithers along, picking up twigs and soil, inspecting the items in turn, then discards them with contempt. Above our heads the tree's branches sway and bobble as if it wants to move and find shelter from the wind.

The poor old tree is very limited in the choices it can make. Most of its growth pattern is pre-programmed before its original seed is germinated. The only events that have happened to it by chance are where the fertile acorn fell and how the climate has subsequently shaped the tree's canopy. If we could communicate with trees we would probably find they were all fatalistic in their outlook because of the few choices they can freely make. That is, they would believe all future events were preordained. Consequently they would need to envisage a reason for this based upon the most important aspect of their existence. I imagine they would worship air, since it sustains and shapes their structures at all times. Perhaps their carbon-based conscience would go one stage further and construct a metaphysical system like that of the early Greeks: the elements earth, air, fire and water would provide an excellent frame of reference for tree activity. Fire may be considered the evil element, the ultimate fear.

A rigid determinist would have taken me to task by now, claiming where the seed fell, and the future climate, could all be determined if enough data were available. In other words, if it were possible to know the cause and effect of every action, the result could be predicted in advance. This can be very nearly true in a laboratory situation, with one item under investigation and all other variables held constant.

Some people believe their future is already documented, waiting for them to flawlessly enact their part in a huge cosmic play. I reject all this outright. My point is although determinism makes for a good argument, it is the practical application of the theory which undermines the idea. For example, one of the world's largest information gatherers is the computer at the London Meteorological Office. Weather data is compiled, correlated and fed to programs inside huge memory-capacity computers. And what do they predict? Weather forecasts that are eighty-five per cent accurate the following day. Good? Perhaps, but not totally correct. In one memorable year a chap from Yorkshire was more accurate using his knowledge of nature as a guide – field law, as it is called. So if we cannot determine tomorrow's weather with confidence, why believe in inevitabilities?

The answer may lie in Western man's affluence. Generally speaking, he is clothed, fed and sheltered. It is far easier for him to believe "whatever will be, will be" rather than to get involved in all that useless philosophising. Drifting along is easy; far easier than trying to alter course and sail into the wind. That action creates turbulence and can be detrimental to the individual's health.

Unlike the tree, human beings have a greater scope for choice: there is no need for people to be deterministic in their outlook. Fatalism is a false idol to worship. In a capitalist democracy, like ours, people have so much choice everything they do is either taxed, needs a licence, or both. It is not as free as you might think.

Although human genetics do not comprehensively dictate our lives, most seeds are not dispatched far from their parent tree. In reality we are each born into a particular social/economic background which guides the rest of our lives in a particular direction. Yet inbuilt in us, additional to our instinctive behaviour, is a portion of memory available for open programming. In computer terminology a random access memory (RAM), a large area of storage set aside for learning, a place to contain our particular culture.

Our culture is taught to us when we are young and, subsequently, it is here, what we teach the young, which largely determines how well they survive. The choice of what to teach them dictates the success or failure of the next generation and ultimately the human race. But who decides the syllabus? In a democracy the people should ... at least, that's the theory. We are approaching the realms of politics now, but first a quick look at another human trait, the reverse psychology phenomenon.

On Saturday morning I am always up and about before the postman arrives. Linda has usually had her cereal, milk and vitamin drops by now and is gaily tripping around waving a piece of buttered toast. Tracy is still in bed; it is her morning off-duty. She is lying there wondering why we give vitamin drops to Linda when we are the ones who are worn out.

The letter-box chatters and Linda points to the hall door and says, "doooo". I gather the letters.

"Take this to Mummy, please," I ask Linda and hand her a brown package, a wool catalogue, I suspect. She obeys without hesitation. Her understanding of vocabulary has increased tremendously and she can now follow most instructions when they are built around a few key words. When she returns I have opened my letter, but with a difference; I am pretending to read the envelope. I ignore its contents (a bank statement) placed beside me on the couch.

"Ooh!" Linda gallops over and collides into my knees. She grabs the envelope, which I do not want, and runs away giggling. Meanwhile I am glancing at my bank finances (in particular the lack of them) and say laconically, "Linda. Bring that back." Naturally she doesn't. She is happy and I have read my bank statement in relative peace.

That old reverse psychology can be so subtle we often do not realise when it is happening. Sometimes we overlook its effects.

Humans have a tendency to create their own philosophies based upon an antithesis of the prevailing feelings and thoughts associated with their formative years. This could be due to the disappointment felt when, as youths, we realise the world is not the utopia we might previously have imagined it to be. There is a desire to change the world without first examining the existing system to see where the changes, if any, are needed. The emphasis is on change without appreciating the stability that existing institutions foster.

This reaction to the established culture is a feature of human evolution which causes it never to stagnate; only pause occasionally. We encountered the tree's analogy to this phenomenon a few chapters ago. The tree grows taller on the side away from the prevailing wind, the sheltered side of the crown. The political winds of our youth buffet our individual trees of life; they shape us; they mould us. Our reactions grow to oppose the force of the ideologies impartially thrust upon us. This is a time to be careful. Do not let the occasional belief grow into sacrosanct hardwood. Always keep those branches of thought flexible. Keep them flexing, too. Otherwise you will be fostering dogma, presumed fact based upon unsubstantiated evidence.

During my lifetime British politics has been dominated by a two-party system. Basically, one preaches socialism and the other unfettered capitalism. These opposing views of society have been labelled left and right wing respectively. This nomenclature became a part of the political terminology after the French Revolution of 1789. Then the format of the French National Assembly was arranged in a semicircle. Those wanting radical changes to favour the people sat on the left wing, while those preferring more established methods were to their right. In Britain, the political divide between the two parties in the mid-sixties has not been, I would say, exceptionally wide when placed in a world context and may look something like this:

```
                    Centre ground
Communism--------------------------------------------------------Fascism
             Labour            Conservative
```

The two main parties are seen relatively close together with communism at the extreme left and fascism to the far right. Nevertheless, the distance between the two major parties has had considerable consequences for this country and what follows is my view of the damage this division has manufactured. It is an essay on politics created out of my own personal experience.

In a survey it was discovered only fifteen per cent of the general public were very interested in politics. This implied the remaining eighty-five per cent were *not* very interested in politics. Hence I imagine that Mr Joe Public has never been to a political meeting and may not know the name of his Member of Parliament (his MP). Joe is not a party member and regards a political party broadcast as an interval in which he can make a cup of tea. In fact, to say he is not interested could well be an understatement. Politics seem beyond his comprehension and control. Events happen to him over which he has no influence; at least, so he thinks. He is a fatalist at heart. Sometimes he never even bothers to vote, having decided a long time ago that a good politician was rarer than an honest second-hand car dealer. Not voting is his idea of a protest against the system, since if he *does* vote he is captured by it and placed into one faction or another. Yet, a lot of people want him to vote and at election time he becomes popular. Then canvassers crawl out of the committee rooms to sample the opinion of this public nonentity. It's a sort of canopy committee in reverse.

Each party runs their campaign with the fervour of a military operation. The candidate's agent acts as the general, organising and planning the battle strategy. His first ploy is to assess the lie of the land.

Every eligible voter is listed, along with his voting number. The list is subdivided so each canvasser (the ground troops) covers a specific terrain as swiftly as possible. He has detailed instructions to follow. First assess the voter's support for their candidate.

Is it (a) non-existent, (b) assured, or (c) indecisive?

If (a) ignore the idiot and move on.

If (b) compliment the person on their choice and move on slowly; inoffensively. Do not become entangled in any doorstep banter. This will lose time and may be counterproductive.

If (c) – an indecisive voter – be polite. Place an asterisk by their name. The general will later dispatch a word merchant of higher ranking to try some gentle persuasion at a later date.

Yes folks, this is your genuine democracy at work. It's the same technique as selling double-glazing.

Back at HQ the campaign organiser (the general) knows exactly the nature of the problem: an indifferent electorate whose hobbies include applied apathy and self-interest. To him, about forty per cent of these "morons" are, well, morons; fifty per cent are of average or slightly higher intelligence. The remainder are so bloody clever they can reduce the candidate to the level of a bewildered schoolboy, which is not good for his image. So the general concentrates on Mr and Mrs Average. The logistics are simple: there are more of them and they are easier to manipulate.

Mr Average does not spend a lot of time reasoning. He did once but it gave him a headache. Now he prefers to listen to and to read what others think. In a sense he is asleep during the day as well as at night. During the day his mind is at rest and during the night his body. Others do his thinking for him – subtly. They seem to comprehend the problems since they often state they know the answers. For example, the socialist knows the answer to the country's problems is to allow the workers to control the means of production. Unfortunately, the capitalist (who is *always* in control) is convinced private ownership and competition between producers serves the country's needs best: two diametrically-opposed beliefs which each party holds to be a truth: two dogmas in conflict.

Election Day arrives and Mr Joe Public has to decide between the two ideologies. You may be disappointed (but not surprised) to hear Joe did not resolve the conflict by visiting his library to research the history of Western political philosophy since Aristotle. His decision is based upon five leaflets, three of them being the election morning "drop". Why three? Oh, there is a third force involved. A party which, we will say, is at the centre of the political spectrum.

The centre party do not believe in extreme polarised solutions. They believe in both socialism and capitalism there are facets which are worth pursuing and those which are not. They are aware good government needs to be a flexible medium able to respond to the winds of change. Their policies seem hard to define, appearing to be a mere blend of what the two main parties preach with conviction. They are called "middle-of-the-road", whereas they really embrace all roads and are not afraid to explore any avenue to seek the best practical (as against ideological) solutions. Nevertheless they remain unpopular with the masses and are regarded with suspicion. Perhaps no one out there can believe sincere politicians really do exist anymore. The masses have become cynical.

Mr Joe Public does not want to get involved with grey areas of debate. He wants to know where he stands. He labels the middle ground "the soft centre" and keeps voting for one of the flanking wings.

It is very hard to muster support for a party that believes in co-operation politics. If Mr Joe Public is working class he may vote socialist because he once overheard (or did he read it in a newspaper?) the socialist's aim is for all to have equal opportunities. He knows, empirically, he is suppressed by the capitalist management at work. He does not want to talk to these opponents and reach an agreement; deep down he wants to smash the bourgeois enemy. He dreams of winning the pools, thinking this will be lucky, and then he can tell his inline management where to redirect their workload.

The manager is generally a capitalist by heart. He wants to be a strong manager. He does not want to consult with his inferiors – his workforce. He wants to dominate them silently and grind them into his schedule until peak efficiency is achieved.

The two drift apart.

Drawing a straight line to illustrate the polarising effect of two party politics is an inadequate representation, understating the divisiveness. A better example will be (you've probably guessed it by now) a pyramid.

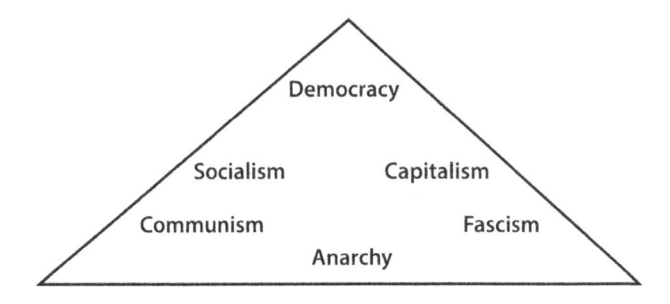

```
                    /\
                   /  \
                  / Democracy \
                 /            \
                / Socialism  Capitalism \
               /                        \
              / Communism        Fascism \
             /_____Anarchy_____\
```

In this diagram, socialism and capitalism are on opposite sides of a pyramid. Democracy is at the peak.

It is hard for a politician on one side of the pyramid to fully appreciate what life is really like on the other side. He probably has no practical knowledge of the conditions which create the opposing point of view.

There is no level common ground between each side, which is why the straight line representation is weaker. Ascending the slope towards a more favourable vantage point is a tiring journey, both physically and mentally. It is far easier to slip down the slope towards a more remote, uncompromising point of view.

At the pyramid base the anarchist welcomes new recruits with open arms. Here democracy is considered a luxury the country cannot afford. The people are felt to be ill fit to decide their own affairs. Practical communism of the Eastern Bloc permits only one party which (theoretically) represents and benefits all of the people. Yet, when the people complain about the lack of farming equipment, material goods and social bankruptcy, military power is mobilised to suppress the unrest. Capitalist subversion is then blamed for usurping the masses.

The other extreme is fascism, usually a military junta, a dictatorship which forbids political debate and replaces it with fear and torture.

Although British politics approaches the apex of the pyramid, it never seems to get there. We call our institutions democratic, but no government in my lifetime has had more than fifty per cent of the available votes. In other words, they have never had a majority support to enact their rigid policies. As a consequence, when their policies flounder, as sometimes happens, support crumbles and shifts towards the opposition.

It seems a terrible waste of energy when great effort is channelled towards fighting the opposing point of view regardless of the merits of the case in hand. Opposition is mustered with fervour to defeat the enemy, be it right or wrong. The division widens; the two main parties slide further down the slope, leaving a vacuum in their place. The idea of a loyal opposition becomes naive.

The British Parliamentary system uses indirect democracy. This means instead of all the people voting on all of the issues (direct democracy) we elect representatives who enter Parliament and speak on our behalf, or so we assume. On some issues, notably the return of capital punishment, our representatives dismiss their constituents' opinions (which have recently been strongly in favour of the restoration of hanging) and vote against the wishes of the very people who put them in office. You cannot get more indirect than that.

Why does the Member of Parliament (MP) behave like this? What makes him tick?

A leading political commentator, who had spent seventeen years reporting the events inside Parliament, had this to say: "Of the six hundred and fifty MPs, probably ten or twenty per cent are normal human beings. The rest are either fanatics or rascals."

He also thought Parliament has had a bad time since the Second World War, the period upon which I, too, reflect.

The MP's loyalties are split. His responsibilities are (in alphabetical order): constituency, national, party and personal – i.e. himself. The MP's order of priorities are (I suggest) nearer to: personal, party, national and constituency. His personal well-being is the main priority; that of his constituents is far less important. For the time being we will accept this; there is an element of self-interest in us all. What I want to highlight is the MP's devotion to party. The leading political commentator continues: "The open-minded quickly join the game of denouncing the Opposition regardless of what they do." (Kept in line by the party whips.) "The Tories using theirs more readily than Labour."

Why should an MP want to adhere to a particular group rather than be free to use his discretion and ethical judgement without paying heed to party prompting?

"MPs have a power mania which they must satisfy regardless of cost. Untruthfulness reigns," we are advised.

By now you will have gathered I have a normal opinion of politicians in keeping with the eighty-five per cent mentioned earlier. One of my reasons for this is their flagrant hypocrisy. For example, if a government were to outlaw the trade union practice of block voting, it could do this by using what is called a three-line whip. In effect, ordering all the government MPs to vote for the measure. Thus they would achieve their objective of banning block voting by creating a block vote within their own ranks.

Nearly every vote taken in Parliament is a block vote: a party vote. An MP is not his own man. He cannot afford to be. It would be detrimental to his career to vote with his conscience.

There has been a lot of fuss lately about unions being compelled to hold secret ballots instead of the traditional show-of-hands means of voting. Yet there has never been a secret ballot in the House of Commons. If there were then perhaps a more accurate representation would emerge, one not necessarily in accord with the Government of the day, but perhaps more representative of the true picture. With no secret ballots in Parliament, the party whips can intimidate wavering MPs to toe the line. Yet this is exactly what union leaders are accused of doing to their members.

British democracy suffers from exactly the same symptoms as all other pyramid hierarchies: divisiveness, inflexibility, all the power focussed at the apex. Policies come from the top down-the-line to affect the lives of millions. To win elections, the policies have to be marketed, sold and advertised like cornflakes. The elector has two main brands to choose from. Each contains different ingredients and both claim to provide beneficial nourishment.

The existing system places an MP's priorities in the wrong order. His first concern is to protect his own box. Next he struggles to ascend the pyramid, hence party devotion. Then he fears the pyramid (the nation) itself may be altered. But lastly he does not pay enough attention to those below on the pyramid base. They are just numbers on a canvasser's clipboard.

A division is apparent here similar to the theoretical/practical split we saw earlier in the factory pyramid. The MP, at the top, never feels the practical effect of the theory he spews forth and implements. He is already financially secure for life. The deepest recession has no effect on his standard of living. The feedback he receives is statistical, not physical; he painlessly remains resolute, loyal to his hardwood convictions. Like the factory manager earlier, he remains consistent, thinking this a virtue, even though he (or she) is consistently wrong.

Meanwhile the poor old cajoled voter suffers the consequences of the political philosophies. He has to cut back on luxury items like coffee, a bottle of wine at the weekend, trips out in the car, new clothes, haircuts and generally all the things that give life a bit more zest and quality.

Can the structure of the British political system be improved? Of course it can, everything can be improved to a certain degree. There *is* an alternative. However, if the improvement is delayed for too long, the masses become disillusioned and turn more towards the extremes.

Below is my idea of the pyramid hierarchy of government:

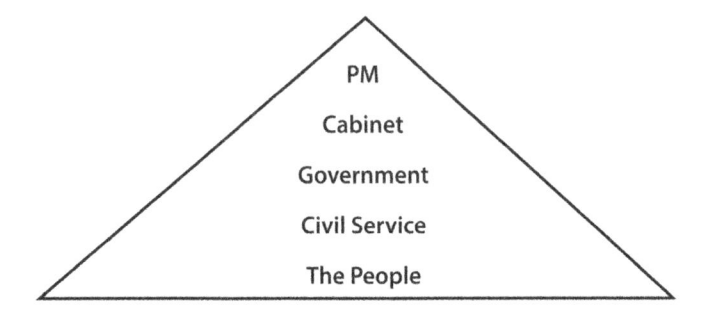

```
        PM
      Cabinet
    Government
   Civil Service
    The People
```

There are a few points to note: firstly the monarch is not mentioned. This is a pity, as the heritage seems wasted in a purely ceremonial role. Granted, one or two of the monarch's ancestors were rascals, yet the monarch still symbolises (for me) a thread of continuity which could enrich society. They could be thought of as the roots which span the evolution of democracy in this country.

The pyramid has no layer to represent the Opposition. If a government has a working majority in the House of Commons, the Opposition is effectively non-existent. Yes, they are allowed to argue and let off steam, but the fact remains they are just there to make up the numbers and any voting will go the Government's way, unless they arrogantly fall foul of a rare unforeseen crisis.

The Government's philosophy is thus fashioned into policies that are pushed through the parliamentary process and left to the Civil Service to enact and apply.

Eventually the policies reach the people.

Remember, under half the people have voted for the Government. The remaining people (the majority) have doubts about the Government's approach and may even wilfully oppose its policies. The pyramid base becomes unstable, forever shifting; it rocks from left to right. The see-saw motion dissipates energy which could be, should be, harnessed. There is a loss of spirit.

Transforming the parliamentary pyramid into a tree structure will be harder than reorganising an individual's tree of life, or transforming a factory framework.

My idea and thesis has been to progress from a knowable and finite system (yourself) to a more complex system (a factory) then on to a huge intricate system of government.

The individual can shape his own tree of life by willpower and effort. The factory tree can be altered by a minority of willing individuals; but transforming the Constitution, the character of a country, demands conscious desires of many millions of people with like minds.

Fashioning a change of attitude towards an objective requires a focal point, an image, a catalyst, a stimulus; a trigger from outside which will reawaken our genetic programming to unleash a wave of co-operation unseen for thousands of years; a tidal wave of friendship and altruism to sweep aside the distrust and disharmony of the twentieth century.

A thesis is needed which will imprint itself on to a majority of minds. We need more people looking for a *new* way.

Now, as I write, there has never been a bigger disconnect between Parliament and the people in modern times. The masses are restless yet trapped by the Constitution. This is not good in a democracy as mature as ours. Only apathy is keeping the peace.

Yet, in the Middle East all Hell has broken loose amongst their peoples who are undergoing an "Arab Spring". Perhaps things aren't so bad here after all? They are fighting for representation, but it will be many years before they reach our level of democracy and are stopped in their cars because the depth of tread on a tyre is less than 1.6mm across three quarters of the tyre width, thus incurring a sixty pound fine and three points on their licence.

Meanwhile, Western democracy has divided countries into internal warring factions, like the Democrats and Republicans in America. Likewise in Europe, division is evident in other major countries. Strong polarisation of thought is everywhere while consensus hides its head. Nowhere has the representation the people need. The wisdom of the crowd is missing.

Branch 13

Linda's Game

Ninety days after the official redundancy notification the pruning of the factory workforce commences. The selection procedure is based upon the "first in–last out" principle whose history I am unsure of. I am also unsure this is the best method to use if the rest of the factory structure is to remain healthy.

When I prune my roses in late autumn, my aim is to remove all the dead wood and most of the old wood. Then the new young shoots can grow unimpeded. A vigorous new plant will hopefully emerge out of the old established rootstock. Factory pruning does not work on these natural principles. The apprentices are the first to suffer.

Right in the middle of the redundancy is a friend of mine trapped in a typical pyramid anomaly. He and I signed on together within weeks of each other. He worked his way up through the ranks, first to chargehand, then to foreman level, crossing the staff–industrial interface gulf. In doing so, he is now the last staff recruit at foreman level. In the meantime I contentedly worked my way up to (unofficially) fourth-best electrician on site; still on the industrial side of course and not quite the best electrician in the world. Now then, because my friend is the *last* staff recruit, he is *on* the redundancy list. (Last in-first out applies to staff, too.) Meanwhile I remain among the senior electrician category; dead wood, perhaps, but as safe as houses.

I think it is fair to say we have both worked relatively hard giving a fair day's work for a fair day's pay. But my colleague's reward for bettering himself could be (to use the vernacular) the silver bullet – one year's redundancy pay. Some encouragement.

The other young growth, on seeing this, makes a vivid mental note not to clamber for ill-timed promotion. Meanwhile, the apprentices are cast adrift, their contracts ended.

In this example, factory pruning removes younger shoots before their potential is realised, leaving older wood behind, which can artificially retard the growth of the structure.

The only hope for those on the hit list is enough volunteers will accept the redundancy terms. Some do so and emigrate. In this way the factory can shed units of labour who have little interest in its wellbeing and this effect may just cancel out the "last in-first out" anomaly.

All this for the sake of efficiency.

Efficiency: it's the latest trend: efficiency in industry; efficiency in the economy. This equates, in crude but very real terms, to fewer jobs. Efficiency is inversely proportional to jobs, or so it seems.

Not long ago when unemployment was below one million people, the country was coasting along and most people had settled into a state of equilibrium with their fellow men and women. To the extreme left and right wing this state of affairs is abhorrent, a complete and utter passive disaster. What they want is unrest. A general election occurred and the right wing of the Tory party attained power. They believed they had a mandate to thrust the country towards greater efficiency. As usual, less than fifty per cent of the electorate supported this idea.

You may recall earlier during a typical day at work I tackled an air-conditioning unit. Different refrigerant pressures inside the unit extract heat from one part of the system and deliver it to another. Getting back to the election result, Tories use a comparable technique to extract wealth from the different levels of society. They ease the tax burden on the rich, ostensibly so the rich will stay (and not emigrate) and invest their extra income in Britain. The poorer people are put under constant financial pressure to help pay for this generosity. The poor have to stay anyway and do not have any spare money to reinvest.

Economic Darwinism reigns; survival of the financially fittest; profit is the resurrected idol.

Heat removal is assisted by two largely differing reservoir levels in a system: hot and cold. Extracting wealth is easier if the gap between rich and poor is widened and this is exactly what happens. The capitalists fashion this imbalance, fearing the consequences financial equilibrium would mean to them.

The economy and industry is savagely pruned for three long hard years. Then the claim was made that the country enjoyed greater efficiency. The truth is only the rich enjoyed it. Industry was hit hardest; even firms with full order books went to the wall. At the end of this period unemployment was above three million. Government blamed a world recession, yet their policies did nothing to alleviate distress; instead they enhanced the depression.

[Much of this sounds so familiar thirty years later in austerity Britain. Political progress has been nil.]

The businesses that did survive apparently became more efficient. Government statistics were arranged to emphasise this aspect. They naturally ignore the inefficiency involved in keeping two million extra people unemployed; people who need ninety per cent of a working wage to survive, yet produced nothing.

Efficiency at all costs created a lot of turbulence. Law and order broke down. For the first time in my lifetime there were street riots in major cities. The Government had pruned too hard; cut off and isolated too much young growth too rapidly. Youth existed in a wind of despair with no hope, nowhere to go and nowhere to grow. The structure of society was being drastically changed in the face of another onslaught from biased, some would say, extreme policies. Any dissenting voices in the

Cabinet were curtly dismissed. The country apparently needed a cure and this was the medicine.

Nowadays medicines come in pleasant flavours. Linda thinks her occasional medicine is a special treat; but the flavour the country was treated to was nasty and hard to swallow; a Victorian vintage; a brimstone and treacle concoction.

Is it possible to apply tree theory to the parliamentary pyramid? Will it create a better kind of democracy, one fed from both ends? Let us examine the existing parliamentary tree, given below:

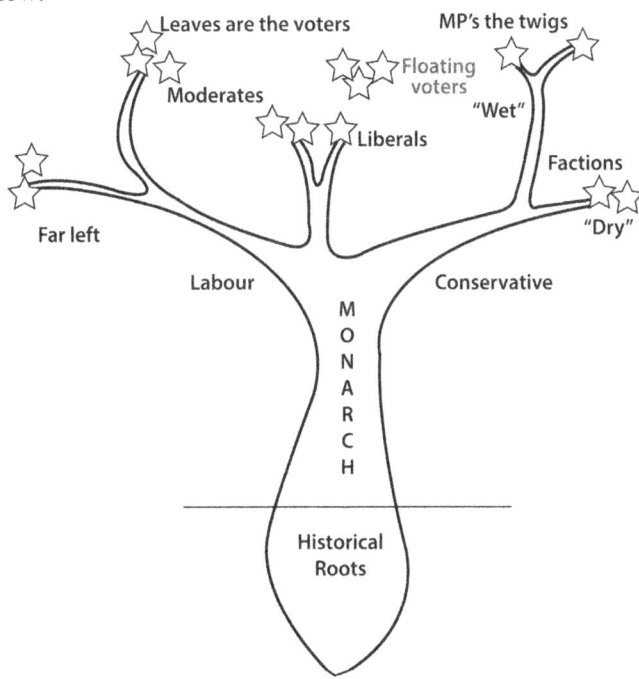

Leaves are the voters — MP's the twigs — Moderates — Floating voters — Liberals — "Wet" — Far left — Factions — "Dry" — Labour — Conservative — MONARCH — Historical Roots

The reigning monarch is the main trunk. No matter how the tree shapes up above, the monarch provides a continuity of culture drawn from historical roots. Remember, all above are subjects of the realm, whether on the left or right branch. They are all part of the country's heritage; tied to it; bound up in it.

The trunk does not waggle the branches; it provides the stability for the rest of the tree to grow upon.

In our parliamentary tree all the waggling is done by the Government branch. It is the larger of the two main branches, united on most issues, but it, too, has factions. Any dissenters in the present Conservative party (Tories) are labelled "wets". The rest are presumably "dry".

The twigs at the end of the branches are the MPs. If one had the time one could work out the connection point for each MP to the sub-branches and hence his association with the rest of his colleagues.

The voters are represented by the leaves. Those who voted in favour of the Government are actively encouraged, being at the canopy end of the Government branch. Other leaves cling to the stunted growth of the Opposition branch. This branch is not (at present) favoured by the prevailing winds. Its resources are gradually withdrawn and channelled towards the Government branch. So the Opposition branch is buffeted at one end and starved of resources from the other.

The little central branch is no more than a stump. Just a few well-meaning pragmatic academics preaching co-operation politics which none of the leaves can fathom.

Swirling around in the middle of the crown are the leaves without any affiliations, ideals, or beliefs: the floating voters.

Inspection of the tree reveals imbalance, hence an unhealthy structure. From the trunk upwards the tree is split into two. One branch controls the rest of the tree, dominating the others. In time views and opinions will change, the wind will alter direction. The healthy branch will wither and the poorer branch will accede to eminence. This continual alteration of resources is considerably debilitating for the tree. Many of the leaves become shaken and disillusioned.

Applying corrective action to this tree structure is an immense task. The pyramid of government (when inverted) needs subtle adjustment and this is hard to achieve against a background of apathy.

A factor which favoured a successful factory transformation was the common aim of the workforce in the sense that they are usually focussing on one product being manufactured. Hence there is a common goal. The political spectrum spans two solutions, a dipole arrangement with two goals at opposite ends; tricky.

I decide to ponder upon the problem for a while. I could visualise individual lives and factory hierarchies based upon natural organic tree structures, but I could not envisage my theories about government renovation being accepted in practice. In particular because the Establishment, who were (in the main) living well now, would see no need for any change.

The time had come to leave the lofty heights of tree theory and come down to earth again.

The year rolls onward. Linda grows more alert, agile, and taller, too. She is becoming a bit of a lovable rascal, especially when she is helping her Dad in the garden.

The sycamore tree is now in full leaf with its flowers starting to form. The oak trees lag behind a little and are not in full leaf yet. The youngest oak unfurls its flowers first; the middle oak follows a week or two later; then the elderly oak (the statesman of the trio) lumbers into life. They seem like a family; baby, mother and father, yet this is a false impression. They are just oak trees of differing ages.

During its lifetime a tree grows slowly at first and then a period of rapid growth follows. Gradually the tree's growth slows down until it gains no more height but thickens around the trunk instead. This growth pattern is applicable to many natural systems, even man himself. There is something (which I haven't yet discovered) limiting the size of nature's designs. Somehow the tree knows when to stop growing. However, one of nature's inventions, man's brain, has evolved a method of overcoming all normal limits: imagination. Man can conceive of a Universe expanding infinitely without limit. Meanwhile, back on Earth, he constantly pushes for growth. Large multinational companies are expected to reap bigger profits year after year. This is what we expect of a successful conglomerate. Would you consider a company that broke even successful? No, you have been conditioned to believe this is failure. Moderation is portrayed as a failing by the advertising media which fuels the fires of demand. Once again we have a situation where man-made concepts (like infinity) have fallen out of line with immutable natural laws. Eventually the exploitation of the Earth's natural resources will reach its ultimate size. Unless we can reach for the stars, expansion on Earth must have finite limits.

Linda's Dad has just cut two lawns. He is at an age when his waist is thickening around the trunk and he is in need of a rest.

I am sitting at the bottom of the garden when a little head peeps around the corner of the garage. The face smiles and says: "Boo!" Linda has come to help me.

"No! Don't touch that! Your fingers ..." We both race towards the lawnmower. I arrive first. Linda likes these chasing games.

"Look. Do something useful," I advise. "Put this grass into your bucket and bring it to me." Might as well try to harness all this youthful energy, rather than waste it. I'll start her tidying the lawn then progress to digging the vegetable plot; that's the theory, anyway.

I put the mower away and return to the bench. Linda has disappeared for the moment while my back was turned. I know she's safely contained behind one of my Dad's gates. I half expect her to reappear, but I can hear next door's children calling her name over the fence. Perhaps she is being amused by them.

The back garden is looking quite tidy after the long hard winter. The lawn is still patchy though and could do with a spring feed.

The fresh, invigorating smell of newly-mown grass wafts towards me on a cool, gentle breeze. I wonder why they don't bottle the aroma and sell it as an air freshener. Perhaps they do?

The tips of the potato haulms are breaking through the soil's surface and the broad beans have bounced back up in the last week.

I still think it is amazing that most of the perennials have survived such a savage winter. I suspect this may be due, in part, to their burial by a morass of decaying leaves – organic insulation. Now the leaves have been removed the plants burst into life again: honeysuckle, pinks, aubrietia: primroses: fuchsia and aquilegia. Even the dried pampas grass, on closer inspection, has little green shoots around its withered base.

In the herb garden the sage and the mint are still thriving. Everything else has gone, but the mint has spread like wildfire and has to be cut back.

In front of the bench is a barren patch where very little grows and Linda's sandpit sits in the middle of this area.

"Boo!" Guess who?

Now what is she up to? She has picked up the bucket again and is filling it with ... my rockery flowers!

"Linda. Don't." One word from me and she does what she likes.

I stand up. Linda squeals with glee and scampers away back up the path. Good game.

I sit down. Linda scuttles back, plucks another handful of aubrietia and turns to look at me. She is smiling, Dad isn't. I rise again and gallop up the path to catch her before her legs can overcome the inertia of her wellies. Aha! Caught her! Time for a lecture. She giggles with glee.

"Don't pull up the plants," I order. "It's naughty. You mustn't do it." I know from experience she understands the message. But Linda is not taking any notice. The adrenalin is flowing and she is full of fun, happy with the new game she has invented. So I try tactic two: change the subject.

"Come and play in the sandpit. That's what your bucket is for."

I lift her bodily to the oasis of sand near the middle of the garden; a miniature Sahara surrounded by its rigid version, concrete. I build a few sandcastles which Linda promptly knocks down. I build a few more.

While Linda is absorbed in her own little world of play, it occurs to me the garden itself is a miniature world of my own making which has coincidental similarities with the real world. The garden is split into three distinct zones. From my vantage point these are: the lawn, the vegetable patch and a potpourri of land in front of me. The real world has similar zones: communism to the East, capitalism to the West and a mixture of cultures in the southern hemisphere.

The southern hemisphere is prone to extremes of climate, drought and monsoon. Huge populations try to eke out an existence from infertile land. The knowledge and resources to feed the world's people and limit their numbers are available today here in the northern hemisphere, but the two superpowers channel their reserves inwardly on defence – military hardware.

The United States' defence budget is above one hundred thousand million dollars a year and the Soviet Union's arms expenditure is slightly greater again. The latter is a considerable drain on the Russian electricians' resources; almost fifteen per cent of his efforts are spent on defence. The British and American electricians each have five per cent of their gross capital earnings spent on defence. Total world military expenditure exceeds five hundred thousand million dollars per annum; staggering.

The balance of power (at the superpower level) upsets any normal altruistic balance individual man would normally show to his fellow man in the Third World.

Deterrence is the name of the game, each side matching the other, covering and countering imaginary scenarios move for move; atomic chess; nuclear keeping-up-with-the-Jones.

More staggering figures emerge so the mind cannot appreciate their implications. United States: nine thousand, three hundred warheads, total explosive power three thousand megatons of TNT: Russia: seven thousand, three hundred warheads, explosive power five thousand megatons. The mind searches for a comparison it can comprehend. The bomb dropped on Hiroshima was equivalent to a "mere" twenty kilotons of TNT. It killed one hundred thousand people. A one megaton explosion can ignite a newspaper at ten miles' distance, which would be awkward if you were half-way through a crossword.

It is beyond a mere electrician to understand the figures. He makes no attempt. This does not worry him unduly, but what does unsettle him slightly (especially during periods of international crisis) is the fact the two sides are not on speaking terms. He finds this ... staggering.

One-third of the world is underfed and the other two-thirds are accumulating vast arsenals, ostensibly for defence. Each side has the destructive power available to destroy the world (as we know it) one hundred times over. To destroy the Earth only once would seem a setback; and so the antithesis is generated. Peace movements emerge, their support drawn from many quarters. Some are fanatics, some unemployed; some build peace camps and cause a nuisance. The newspapers highlight these elements and forget above all these people are mainly concerned about mankind's future survival. This is in itself no crime, yet anxiety over defence is portrayed as a subversive up-swell.

The ultimate conflict in the modern world is a political affair between Western capitalism and communism of the Eastern

Bloc. If these two forces collide head-on, in a game of nuclear conkers, nothing worth having would survive. Not even the trees.

Can we expect to resolve the final conflict of nuclear war? Is there any reason for optimism about our long-term future?

On the bench, in the garden, the electrician sits and thinks and analyses. Linda is playing in the sandpit, fittingly placed where the North African Sahara Desert would be in my miniature garden-world. The electrician's body remains on the bench in the garden while his subconscious mind floats upwards on an astral excursion. From inside his head he looks down on his body and on a little girl in a sandpit and the back garden of a bungalow. Then he visualises a map of the estate, then a plan of the village, as if being carried aloft beneath some silent space shuttle. Now he has reached a great height. He pictures a green-blue sphere dappled with white; serene, silent. And there he floats.

Beneath him the land mass unfolds and he feels physically interposed between two mighty forces, the meat in the nuclear sandwich; Russia on his right and the United States on his left. He pauses, waiting, wondering if destiny will strike today, now.

The image fragments. An invisible force (like gravity) draws him down back to Malkuth – the elemental sphere of the Earth; back to his homeland, Britain, geographically centred betwixt the superpowers. On each side different cultures, different systems; but he knows the divisions are false.

On both sides the land is made of the same stuff. Above the land it rains, snows, the Sun shines and the wind blows, and different areas are harnessed and altered to suit various needs, just like in the garden. Yet beneath the surface turbulence of political rhetoric the underlying soil is common to all areas. Strangely though (and in each system in either side) the common people seem to feel this unifying facet more than the trunks of the tree organisations. The people feel it more than the politicians. I see this as the only glimmer of hope.

Suddenly my composure is shattered. Linda throws a handful of sand in my direction and scampers away. Too late! She grabs more aubrietia from the rockery and runs up the path – the path between America and Russia. I'm in pursuit. The plant is in her hand. Then I arrive and tap the back of her hand firmly.

She pauses, shocked I expect, then cries.

Tracy comes out to see what the matter is. I explain and Linda returns inside to calm down and have her tea.

I return to the bench.

Sigh.

Misunderstandings lead to conflict and some conflicts are settled by force.

I study the garden again. Pessimism swamps over me.

I wonder where will be the best place for my nuclear shelter.

Thirteen is reputed to be an unlucky number. In *Branch 13* all the bad things that happened three decades ago have now cycled around again. Redundancies are announced at my old factory. Youth unemployment is high. We are back in deep recession. The gap between rich and poor widens yet again. Established business and high street shopping chains, with memorable brand names, go to the wall. Austerity is the new nasty medicine the people must take as punishment for problems again not of their own making.

The winds of change have engulfed the planet as a whole: with earthquakes and tsunamis and nuclear meltdowns: the Arab uprising stirs the Middle East with war and discontent. However, the Cold War is over, though there are still more than adequate warheads readily available on either side for a nuclear winter. And nuclear weapon technology filters out to rogue states that have to be coerced into any meaningful dialogue.

I try to be upbeat, but everywhere I look the reality makes me wonder if *Homo sapiens* (wise man) is about to outstay his planetary welcome.

Branch 14

Oak Lightning

One of the many advantages of tree theory is this: when confronted by an impasse, instead of plunging to the depths of despair, you seek the solution in nature. To overcome a difficulty requires no more than an unhurried visit to the nearest field. With any luck it will be surrounded by trees to study. I find this a very novel and unique notion. With all other structures you are mainly dealing with geometric abstractions, or, a man-made model. But when you want to investigate tree structures you can observe actual life forms whose basic shape have evolved over millennia long before that of modern man.

Some time ago I constructed the parliamentary tree structure and noted the trunk split into two main branches. Also, the world itself is politically split, bilaterally divided. Frankly, I can see no answer to this division. Traditional theological ideals seemed to lack the small print necessary for practical usage, making them inappropriate as a source of solutions. At the very least I sought a solution which appeared rational. So I decided to let matters rest for a while pending a moment of insight or inspiration.

It was one of the warmest months of May on record. Elsewhere Argentina invaded the Falkland Islands. At the time of this announcement Mr Joe Public imagines these islands lie off the coast of Scotland and not in the South Atlantic. He's never heard of them before and can't understand what's going on. Government cutbacks reduced the naval presence around the islands and the Argentinians stepped in. The British Government sends a task force to reclaim their territory on the other side of the world.

In an attempt to avoid armed conflict the Pope visits Britain, then Argentina. He proclaims "modern warfare is totally unacceptable as a means of settling differences of opinion between nations." However, both peoples (each containing a large Catholic content) seem to accept the inevitability of war in this case. Perhaps the Pope meant modern warfare is a tragic way of settling disputes. His statement, like religion itself, holds an ideal up to the mind. Most religions are founded on good moral and ethical ideals. The trouble starts when they are applied to the inferior human species. For religion to be successful all humans would have to be gods and God would have to be human to appreciate their efforts. Religion could be considered a result of that old reverse psychology again: man's reflections on his forlorn plight subtly provide a solution embodying hope. Who knows? Perhaps we are just actors on a spherical stage playing out our lives to await acclaim or notoriety after the final curtain call. But I like to think we are part of an elaborate adventure game being played out by the Universe and the solution is down here amongst us, disguised, camouflaged, so only those putting in the effort actually pass the test and find an answer; perhaps *the* answer. Perhaps we cannot see the wood for the trees.

June sweeps in with a heatwave. The days follow a similar pattern: a hazy start, a hot unbearable middle, then a dull evening with the occasional thunderstorm. There was a storm last night followed by an earth-quenching cloudburst. It was the first night for weeks when there was no need to water the garden. I was beginning to wonder when God would be having his turn again.

Next morning the alarm clock clanks as usual, but on drawing back the curtains a strange sight greets my eyes. Beyond the hedgerow where the field should be, misty water vapour glows brightly; dazzling. It's like looking into a huge fluorescent light tube stretching out along the horizon.

One advantage of getting up early before anyone else is the panoramas you witness others only read about in books.

I dress, wash and make the girls a cup of tea. I rush my breakfast and don my coat, eager to experience the strange lighting conditions outside.

"I'm going to work now," I tell Linda, who is now in bed with her mum. "See you tonight."

"Byee," she says.

"Byeee," Tracy mimics.

"Bye," I reply.

The debris of a thousand oak flowers is strewn across the drive, dislodged by last night's downpour. My feet squelch on the pulpy morass. I will have to sweep the drive soon, before this mess is carried into the house underfoot.

In the last few months the journey over the hill has taken on a different aspect. Now the gardens have seen some attention it is quite easy to divide them into two categories: those owned by pyramid-people and those cared for by tree-people, though the distinction is not decisive in all cases. Sometimes a person makes a gallant attempt at creating a tidy, orderly, yet spontaneous garden and then parks three cars on the drive (two of them rusty) which ruins the whole effect.

Permit me to describe the *new* journey.

I close the drive gate behind me, to keep any stray dogs out of the garden, and wander up the estate with not a soul in sight. The estate was built in two phases. The lower half is filled with homely bungalows, mine being the last one built. The same builder started the second phase but ran out of finance and went broke. Whoever took over crammed as many "conventional" semi-detached houses as possible into the remaining land at the top of the site.

House style and location inevitably influence the occupants and their attitudes. In my mind an association has been established between types of dwellings and the kind of occupants I imagine live therein. A cottage conjures up a mental image of a homely couple; a mellow feeling seems to associate with this picture, too. A block of flats gives bad feelings: harsh environment, angry people and decay. A semi-detached house with a little garden seems unappealing. It may not to you.

I am convinced a good architect could take the land area of our estate and thoughtfully design upon it the same number of dwellings with more appeal added. Simple things like an archway here, or an alcove there, an "L"-shaped room (uncommon?), a feature fireplace, or a balcony; something the occupants can perhaps add to, highlight or adapt. Not just the standard box-box-triangle: ground floor, first floor and roof arrangement which is totally uninspiring. Only the contained area surrounding the modern dwelling is really free for expression. Let's look at some gardens.

Just then I nearly bump into a sleek sports car carelessly parked with two wheels half on the path. The adjacent drive is empty and the garden unattended. Pyramid people 1: tree people nil. I stroll onwards past another ten houses in the pyramid class: scruffy lawns, paper on the paths: rusty cars dappled with primer yet never finished. Oil patches and cigarette ends stain the cement drives.

The corner plot has a nice tidy garden but the remnants of a smashed milk bottle lie glinting on the path and the road by the gate. The glass has been there for weeks. Pyramid people 12: tree people still nil. Perhaps I am setting the standard too high.

I leave the estate behind and head into open country. I am fascinated by the glowing white heat haze on either side of the road; so intrigued I step on a dollop of dog-dropping. Yuk!

While I am wiping my shoe on the grass verge I notice there is an oak sapling trying to escape from the hawthorn hedge. This simple observation fills me with a strange feeling of hope; for if man ever does obliterate himself, along with his horizontal hedge-cutting machines, these dormant giants may once again dwarf the hedgerows and recolonize the land.

By the time I reach the hilltop I am lathered with sweat and my legs feel heavy. The high humidity is having an adverse effect upon my metabolism.

The top of the electricity pylon peeps out of the mist and somewhere to my right a cow emits an unearthly groan. There is a sweet smell of silage in the air.

Now it is all downhill – down into suburbia to inspect and classify more dwellings. For a change, and thinking it a little shorter, I forsake the telephone-trench-topped path and cross the road, cutting the corner off the long gentle curve.

Here is another garden next to a large unattended plot of land. The front lawn is neat and tidy and mottled with black juicy slugs overflowing from the adjacent wilderness. There are dozens of them glinting slimily in the white sunlight. Nevertheless, the garden still appears to be quite appealing and I award this unknown gardener the secret honour of being a tree-person while they probably still slumber in a cosy bed.

I continue down the long hill totting up the running total as I go.

One garden fails because there is a bin bag full of rubbish by the gates. Another good attempt falters when I spot a five-gallon drum concealed under a hedge fanning out over the boundary wall. To get a pass mark requires you must respect and tend to the areas outside of your own little box, not just inside your perimeter fence.

The time comes to nominate the supreme tree-gardener of the day. It is easy to select a winner. His garden is not as perfect as some, neither is it an unkempt wilderness. But outside his garden is a grass verge between the public footpath and his fence. This area has been tidied and the grass is cut short and is yellowy in colour. Here is the *coup de grace*; a wooden post topped by a hand painted sign which says: "Dog owners please note. THIS IS NOT A DOG LOO."

Bravo supreme tree-gardener of the day! I salute and commend you anonymously.

By the time I reach the bus stop the score is about even between the pyramid and tree-people. Perhaps I have been observing a symptom of a much more fundamental division: capitalist-person (hunter) and socialist-person (farmer).

The game was not an accurate survey, just a means to pass the time while I walk a mile. For more accurate tests we will have to turn to statistics. This is the science of counting. Usually its application is to uncover trends which may at first sight seem random, or not readily apparent.

For example, you might wish to know the most common number of people in a family unit. You would have to specify which members of a family to include, say parents and children. Having set the conditions, or limits of the test, you could now commence counting. Family one: three members, family two: four members, etc. You would have to count a lot of families (maybe millions) for a full analysis and this might prove impractical. Hence you could limit the sample size and decide how best to select the sample. This could be on a regional basis, or a random basis, or an age-group basis. And after all this sampling and counting you may find one thousand family units contained three members, five thousand families had four members and three thousand families had five members. Therefore, the most common family unit (in our sample) is one with four members and that is that.

It does not mean four in a family is best, or only good families have four members. However, we can assume the most frequent family size (four members) is just so because present conditions favour four in a family unit. Hence we can infer (at present) four is preferable to three or five members in a family unit (in this example).

Now let us turn our attention back to the parliamentary tree and its division into two main branches. My book on the British

Constitution states the two-party system produces strong governments. I wonder how many branches are most common in natural tree structures. How have real trees behaved, grown and adapted? Some counting is needed to provide the answer. But first an exploratory mission is needed to get a feel for the problem; to examine the sample available.

The box on my bicycle is loaded up with sandwiches and a pint of banana-flavoured milk. It is my lunch hour at work and I intend to explore the hedgerows again. As mentioned earlier, tree theory is a good therapy in itself.

The weather is hot and humid and the sky bright and hazy on the eyes. It is cooler when pedalling along; the rush of fresh air in my face feels exhilarating. Soon I am outside the factory and winding my way along narrow lanes last trodden when covered in snow.

The trees are in full leaf and I have to ride slowly to see how many large branches emanate from their shady trunks. I am not sampling yet; first I have to specify what kind of trees to include, their size and location. I casually continue the observations until I reach my destination, a wooden gate with a "Public Footpath" sign nearby. I unpack my lunch, lock the bicycle to the gate, and climb over the stile beneath the sign.

My thoughts are of food! I can feel pangs of hunger twitching my stomach. I sit on a grassy bank on one side of the track that is the public footpath. The ground is damp but I have no choice but to grin and bear it. No sooner am I seated when a herd of young cows trot over to the fence opposite. They stare and "mooo" in remarkable unison. There is a thin-looking piece of barbed wire between them and me. Unperturbed, I eat my meat-spread sandwiches and drink the pint of milk, a meal based mainly upon the cow. It is no wonder they are moaning at me.

Diagonally opposite, to my right, are two trees, one an oak and the other an ash. Remember, there are three basic shapes: maiden, pollard and coppice, and only the first two will concern us. The ash has the maiden shape; it has grown in a natural upright manner with all the side branches originating out of the central trunk which tapers from root to tree top. The oak tree has a pollard style; the trunk diverges into strong branches at a mid-height point, then branches radiate out to form the crown.

The ash is the theoretically ideal tree shape, the tree that would result given suitable conditions and favourable circumstances. A religious tree: all ideas (branches) emanate from one central theme.

The oak is the practical tree. It started life with the same genetic ideals as the ash but was less fortunate than its neighbour. For reasons we can only guess at, it did not realise perfection. Most oak trees do not and are usually of the pollard style. In a sense, the pollard style is more natural for the oak because it is more prevalent. The oak tree in question faces into the prevailing wind, hence sheltering its neighbour the ash. But there are many other factors which can affect a tree's shape: frost damage, bird damage, squirrels, drought and the weight of snow. As soon as the terminal bud is broken branching occurs lower down the structure and the perfect symmetry is broken.

The oak is top of the British list for supporting insects. It can sustain two hundred and eighty four different insect species and support larger species, too. Its fruit (the acorn) attracts magpies, jays, squirrels and pigeons; all are big enough to cause twig damage. The magpies are frequent visitors among the trees bordering my drive. The jay, in all its mellow hues, is a less frequent visitor. The squirrels I have seen only three times in six years; pigeons are becoming more common though.

The maiden-style tree is the perfect tree structure which is the main reason I am not going to study it. Perfection is rare; mature maidens are fairly rare. What I want to study is a structure which is common, mature and has obviously been weathered. I can see no point in praising the merits of some ethereal philosophical system which, though theoretically (or theologically) complete, cannot be objectively assessed. I dismiss perfection with the same disdain I lavish on the self-confessed expert. Let us not waste our precious time and efforts craving for a system that cannot be attained; like a tree structure with one central idea all others adhere to and are subservient to. This is the stuff of stagnation. Perfection, if achieved, is the end of progress. For then there will be no need for betterment. Is the worm perfect?

I want my future to have a diversity of choice. But the diversity must evolve from common rootstock and then spread into different areas, not be blindly staked to one central theme, but venturing, slowly and purposefully, into new areas of space and ideas.

I decide to apply my statistics to oak trees, a favourable species, I feel. They grow in relative abundance, punctuating the hedgerows at regular intervals both near my home and near where I work. Hence a good sized representative sample is available for study.

With lunch completed I move onwards and arrive at a field set aside for winter cattle feed, a field of grass land fertilised and nurtured to sustain the herd. When the grass stops growing, in late summer, it will be cropped and stored as green, sweet-smelling silage. Mature oak trees line the hawthorn boundary of the field. I climb over the gate and start strolling around the perimeter.

The leafy domes cast large shadows, making it hard to see the diverging main branches. Winter may have been a better time for branch counting, but not quite as pleasant as on this warm summer's day.

I have to decide which trees to include in the sample. Judging by the nearby oak trees, anything above forty feet high looks pretty mature. I let this height determine a tree's inclusion or otherwise. Any tree around this height is passed; those noticeably shorter fail and are not included in the sample.

All maiden oak trees are rejected also, but will be noted to see what proportion of the sample they form. Perhaps there will be few for man-made reasons. Good maiden oaks made ideal beams for building a solid farmhouse roof. Hence the tree population has already been affected by our forefathers. They physically eliminated and used the single trunk trees, which I have theoretically rejected in any case.

Now I know the kind of tree to look for all that remains to be done is to arrange a branch count. Then we may discover how many main branches diverge from a mature tree structure in reality. I am only looking for a guide using very mild statistics. I wonder what nature will reveal; perhaps I am in for a surprise.

That evening another thunderstorm rumbles over the horizon. Our small car is parked on the drive when I arrive home from work. I decide it may be a good idea to put it in the garage. Anything left underneath the trees at this time of year gets splattered with the husks of oak flowers. They wash off the car paintwork quite easily, but penetrate the ventilation grating and blow about inside the car when in motion.

The night is still and clammy. Breathing is a conscious effort. The air entering my lungs is as warm and moist as that exhaled and my body isn't too sure if it is working properly. It's having difficulty detecting the air exchange.

As I walk back up the drive thunder rumbles and resonates in the distant hills. A warm breeze begins to stir the trees. Hot air is being drawn into the heart of the storm. Here it rises rapidly. The warm moisture condenses and cools and ice crystals are formed. The ice particles coagulate into hailstones which fall back under their own weight into the central thermal core of the storm cloud. The hot rising air aspires to blow them upwards again, building more crystals of ice around the hailstones in a circulatory process. And all this is taking place between the negatively-charged Earth and the ionosphere at a potential of four hundred thousand volts; the Earth−ionosphere system forms a huge concentric capacitor. Each circulatory thunder cell generates an immense electrical potential which draws ion tracers out of tall structures such as trees. When the potential is great enough, and a route is traced out, the main discharge blasts through the atmosphere, ionising a blinding silver-white path in its wake. The surrounding air is forced to expand rapidly, creating the explosive sound of thunder. A good cell can generate a lightning strike every thirty seconds and a good storm cloud may have several cells. Eventually the ice crystals become so heavy the thermal up-draught can no longer elevate them. They fall Earthward, sometimes as hailstones. More often they melt on descent and the attendant downpour dissipates the storm's ferocity in a quenching cloudburst.

I watch the dark mass of cloud approaching. Lightning flickers across the sky and the wind gains momentum. The trees above hiss now, as if passing an urgent message to each other: "the god of rain and fire is near."

I run inside to relative safety.

The storm is upon us, not quite overhead, but jolly close. Rain plummets down from the sky, bouncing off the road, penetrating the absorbent earth, cascading off the roof in and out of saturated gutters, then waterfalling in front of the window pane.

"Come and look at the rain, Linda." I lift her up, though she is unimpressed and squirms in my arms. I place her down again.

Lightning flashes again, closely followed by the crackle and roar of thunder as the atmosphere is wrenched apart by another fierce ground strike. Linda looks up with questioning eyes. Tracy tells her the clouds are bumping together, which I think is a pretty good explanation.

The tree that suffers most from lightning strikes is our friend the oak. Its grooved bark is a bad conductor and when it is struck the tree often explodes as the inner sapwood turns to plasma. Old foresters maintain, in a beech wood with just one or two oaks, it is nearly always the oak tree that suffers the most. Scientific research supports this observation. The beech has smooth bark which allows water to steam down it evenly. This makes a good conductor and if the beech is struck little internal damage is done. Incredible forces are at work outside, but there will still be plenty of trees left to count tomorrow.

Unfortunately, the next day it rains and also on the following day. It is all part of life's rich variety, a complex blend of random variables ringing the changes and making tomorrow less predictable. This uncertainty can throw a spanner into the best laid plans. It seems wise to probe ahead with caution, waiting for the right moment to act. And so I wait and wonder and find this enforced inaction quite pleasant, even relaxing.

Here's how to determine the age of a hedgerow.

Count the number of woody species in a 30 metre length; like hawthorn, beech, oak, black thorn, elderberry and wild rose. Then multiply the number of different types (of trees and bushes) by one hundred. Of course the trees will probably be hewn horizontally, along with the rest of the hedge.

So in a hedge with hawthorn, holly, beech and oak (four types) you can infer the hedge is four hundred years old.

However, the modern hedge planting technique is to include a variety of types. In the past the original hawthorn hedges took decades for other species to gain a foothold. So it may not be as easy as it looks.

Now the climate is warmer. Summer approaches and the land turns a lush green.

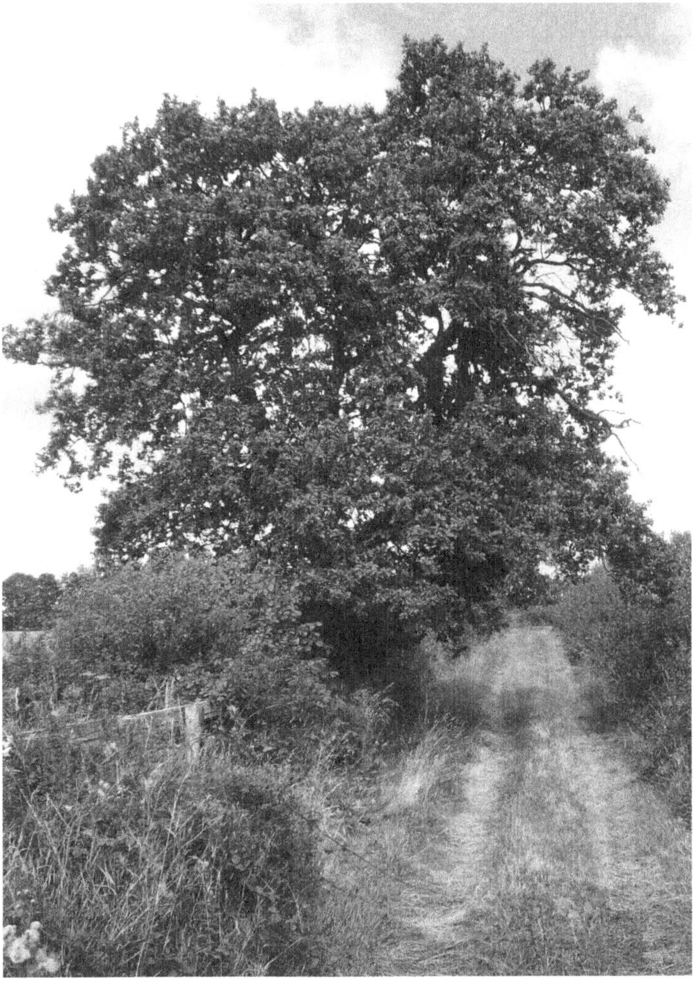

Branch 15

Greek Excursion

There is one thing that has to be planned well in advance: our annual summer holiday.

On Friday night, when an excited little girl has eventually fallen asleep, Tracy packs our suitcases and I load them into the car. Tree theory is abandoned; all serious thinking is switched off and stored in a portion of memory labelled "in abeyance".

The following morning we set off for the seaside and the Sun. It rains again, but only during the lengthy journey. By the time we arrive the Sun is shining, the sky is blue and the beach looks inviting.

Swimming trunks and towels are hastily unpacked. Linda's bucket and spade, ball and small plastic watering can are all pushed into a carrier bag, and we walk down to the beach. The rest of our luggage is now safely locked in a six-berth caravan which will be our base for the week.

The beach is set in a bay contained by rocky outcrops on each flank. The edge of the Atlantic Ocean, at low tide, gently laps the sloping sand; a ripple tide. To reach the water's edge we totter over stony ground. Linda does not enjoy this terrain. She puts her hands above her head and waggles them, a signal which means, "I want a carry". I lift her up and we crunch over pebbly ground to reach the nice smooth sand, freshly washed by the ebbing sea.

I try to put Linda down but she is unsure about her surroundings, so Tracy takes over while I remove my sandals to go for a paddle. Linda watches as the sea froths around my shins. Her curiosity is aroused. She wriggles free from Tracy's arms and is lowered on to the sand. We leave her inexpensive play shoes on (in case any sharp objects lie hidden beneath the sand) and she toddles towards me.

"Come on, Linda." I offer encouragement and hold out a hand. "Come and feel the sea."

She plucks up courage and waggles towards the tide-line; then the seventh wave arrives, larger than the rest, and chases her back to Tracy. The bubbling water recedes and she advances towards me again. She is close now so I grab her hand and she holds on tightly, eagerly.

The sea is cold. My feet are numb already; now Linda gets her feet wet. A wave rushes past her; she loses her balance so I hold on tight. Another wave has the same effect. At first I think the sand is being washed away beneath her feet, but on further consideration I decide the imbalance occurs when her vision perceives the relative motion of the wave passing her feet. Her automatic response is to compensate for the apparent motion by leaning, as if she is running and the waves are standing still. I stare down at the next wave and feel the same effect myself, yet when I raise my head the two headlands border the periphery of my vision, making the scene stable again. Down at Linda's level the effect must be more pronounced.

Tracy has found a large seashell. She calls to Linda, who immediately releases my hand and dashes ahead of the next wave towards her Mum. Oh dear, she's fallen. The wave rolls up the beach just reaching Linda's toes, then Tracy sweeps her up and away to safety.

An hour soon passes by and the sea starts to reclaim the beach, forcing us back towards the pebbles and stones. Linda's outfit is damp and her hands feel cold, so we decide to return to base camp for refreshments and a change of clothing. We wander up the pebbly beach to the stony area. Linda gives the carry signal and I pick her up. Her wet play-shoes form two damp patches on my shirt at waist level.

After a brief rest in the caravan we visit the onsite supermarket to gather supplies. Tracy reconnoitres the food section while Linda and I inspect the adjoining souvenir shop. The shelves are packed with all sorts of holiday-oriented items, everything from windbreaks to flags for the top of sandcastles, and all are made in the Far East in the vicinity of Japan. Linda strains at her safety harness trying to grab the toys, but I'm not letting her loose in here on the first day and keep a tight rein, literally. I buy two magazines, one for Tracy and the other a computer glossy for myself. Linda cuddles a small furry toy duck, a cute-looking thing with a price tag of three pounds fifty. It's not that cute. The shipping charges from Japan were probably expensive – three pounds? I put it back and divert her attention elsewhere.

Something then catches my eye. Tracy is waving to me at the cashier's desk. She wants something. What?

"Money." Oh.

While my attention is diverted Linda has grabbed a toy boat. There is no time to prise it out of her clutches now, so I guide her to the cash desk fumbling in my pocket for a five pound note as we go.

"And this." I put the toy boat on the cashier's desk. She rings up the price on her till, takes my five pound note and gives me little change and a look that says "bloody tourist."

On the way back to the caravan we stop at the children's play park. The main attractions are an assortment of fibreglass and steel animals mounted on large springs. Over at the far end is a combination slide and climbing frame, a bit too advanced for Linda to tackle on her own.

Linda is let loose and she charges towards a larger-than-life pelican mounted on a huge lorry suspension spring. Her arms are raised in the air and I lift her on to the saddle. It is a full-time job keeping her safe, but while she is gently rocking to and fro I am quietly examining the design. (It's an occupational hazard. I will probably check the caravan electrics tonight!) The lorry spring suspension I find particularly novel. It permits three hundred and sixty degree rocking action with a small degree of bouncing.

Linda tilts to one side to dismount. She assumes I'll catch her and, to date, I always have.

The next ride is a fibreglass aeroplane on two big springs; a miniature jump-jet perhaps. She stays on this for…oh…at least twenty seconds then is away again heading towards a gaily coloured bumble bee. A different suspension system here, it only rocks in one plane, but Linda is too light to rock it anyway, so I apply a bit of oscillatory motion to the stinging end.

Now we aim towards the slide. There are other children on the framework and when they reach the bottom I hoist Linda halfway up the slide and let go. To my surprise she stays still. Her feet are wedged apart against the sides and she is holding on tightly to the hand rail.

"Let go," I advise and she winsomely obliges.

Down she slips, without enough momentum to shoot off the bottom edge.

"Maw," Linda says. Okay, let's do some more!

She would be here all night if we let her.

One more circuit of the park, then we drag her away. The seat of her pants is filthy; the slide is nice and clean though.

Back at the caravan Linda reluctantly goes to bed. Eventually her metabolism tires and she sighs into an automatic breathing rhythm.

Peace at last.

Tracy and I settle down to a glass of wine and pâté on fresh French loaves.

The following day the Sun is shining again so we unanimously decide to spend the day on the beach. Tracy and I have learned, from past experience, to sample the Sun religiously, in case it decides to disappear for the rest of the week.

After breakfast I go to buy a paper and fresh supplies of milk. Linda joins me in her buggy, giving Tracy a chance to tidy the breakfast table and organise the beach bag. It is evident on our buggy journey that the tide is quite high at present. We gather supplies from the site shop, then venture along the driveway which runs parallel to the remaining strip of beach.

On closer inspection the tide has been quite high indeed. The tide-line, marked by a dark trail of seaweed, is within twenty feet of the sandbank protecting the promenade along the bay's perimeter. Now the waves are retreating into the ocean in a fluid series of pulsating white foam which hisses and crackles on pebbles below.

We report back to Tracy, gather our equipment and wander down the path towards the sunny bay filled by a bright blue sea. There is a beached dinghy where the path ends and Linda insists on sitting inside it and will not come out until I take her photograph. Now we select a suitable squat for the day.

The good sand still has the sea on top of it, so we make our camp on the pebbly shoreline. The windbreak is coaxed into the ground using a large stone. Other debris is scattered to produce a level surface where we can place our picnic blanket. Tracy erects her chair and Linda throws damp sand into the sky.

By mid-morning the beach is still fairly quiet, a testament to the depth of the economic recession. Five years ago the beach would have been crowded. Now the windbreaks and deckchairs are spaced well apart. People throw frisbees, children brave the water, white bodies turn pink and other bodies catch frisbees. I build a sandcastle and Linda knocks it down. Again she throws sand skyward and it lands over her Mum, who has just finished applying sun lotion and is less than pleased. But Linda's ability to concentrate is below one on the Richter scale and after ten minutes of applied sandcastle destruction she starts to roam.

Her first idea is to lay claim to our neighbour's beach ball, presumably because it is larger than her own, or perhaps just because it isn't her own and therefore new. I am unconcerned about this as little girls (as you probably know) are made from sugar and spice and all things nice and can also get away with murder.

Linda picks the ball up and the young boy who owns it leaves his construction project and runs over.

"Give the ball back, Linda," I instruct.

"Aw. Isn't she nice," says the boy's mother. The boy's dad remains aloof. I imagine the father is worried about his territory being invaded – the churned sandy area around his windbreak. This seems to me a fundamental paradox of mankind: on the one hand man wants to be sociable, but on the other he is cautious and afraid of his neighbour. This atmosphere of mistrust could easily be extrapolated from two windbreaks on a beach to two superpowers on a planet.

I begin to feel uneasy, as if we are imposing on our neighbour's holiday. I coax Linda back to the vicinity of our windbreak and get her to play with her own beach ball.

It is a strange situation; the young of the species are gregarious and fun-loving while the adult creature is often withdrawn and disgruntled. The adults may have something to relearn from the young. Of course, a society based upon fun values alone would be utter chaos. Linda's apparent belief everything she encounters has been made for her to experience contains a degree of truth. For the sea, the sand, the sky, the beach ball, you and I, are all made from the elements of the Earth. The Earth has, in a sense, manufactured us. Now we can choose whether to destroy it or turn it into a paradise. Can you envisage, as you sit reading this page, the atoms in your body are in contact with the chair you sit on and the air around you; which in turn cascades outward contacting every other atom in the world? Can you picture yourself at the centre of a concentric sphere from which your presence radiates out to the entire planet? A human pebble dropped into a pool on the edge of the

Universe. To every action there is a reaction. The smallest thing you do has influence on the sequence of events that follows. An apparently trivial action, say picking a nail out of the gutter, could prevent an accident which saves a life. Leaving a nail in a gutter could kill someone.

Linda appears to be able to feel this effect. The trouble is children not only want to explore this planet-of-play, they also want to keep everything for themselves. The adults have to teach them to share and often this example is not readily forthcoming. The adult is reticent and needs time, or beer, to loosen up. Perhaps, on a beach like this, it would help if windbreaks carried a message, like on some T-shirts only bigger. What a communication bonanza. Let me think. What would a humble skilled manual worker and his family have on their windbreak?

Jeremy, Tracy and Linda
SMW, Teacher, Rascal
Have a nice day

Imagine a day on the beach now, with all the gaily coloured windbreaks to read. The whole atmosphere would be different; more light-hearted.

Chris and Vicky
Newlyweds
Do not disturb

Fair enough.

Sue and Diana
Single: Caravan 69

Hmmm...

Milburn and friend
Mind our frisbee

Infernal things.
Here is an interesting one... ***WHY?***
Good question.

Meanwhile, back on the real beach, Linda is now heading towards two senior citizens who sit quietly staring out at the horizon. They do not have a windbreak and are sheltered behind a large tree stump that has been bleached white and washed ashore by the salty sea. I wonder what has caught Linda's attention.

It was the tree stump; Linda is going to climb it. The two elderly spectators are mildly amused, but their expressions hint at a deeper understanding as if they have lived it all before. Calmly their attention returns to the horizon. Now *I* wonder "Why?"

Linda descends rapidly; she lands with a bump as her bottom contacts the beach. The elderly couple turn and smile. I jog over to rescue her and carry her back to Mum. Soothing words mend her aching pride. Tracy offers to buy some ice creams and the two girls wander away. As soon as they are gone I, too, find myself staring out to sea.

Watching the sea is fascinating for a landlubber. Living inland deprives one of the spectacle of constantly changing tides. You can sit and watch waves all day and never see two exactly the same. The sea is always on the move, yet there is one thing which does remain constant and geometric – the horizon, a mystical straight line which can never be touched.

The founding father of geometry was a Greek named Pythagoras, who was born and lived (for a while) on the island of Samos. Naturally the sea was never far away and the few main geometric manifestations of nature were easily observable in the clear air: the straight horizon, the round Sun, the changing face and curvature of the Moon. Perhaps contemplation of these three things led him to wonder about the relationship between them, the angles through which they travelled, the length of the shadows as the day progressed.

Geometry needed hours and hours of patient pondering before any solid ideas were formed, time which the island situation and the clear skies of the Mediterranean Sea fostered. I doubt if geometry could have such a healthy infancy inland away from the full effect of these powerful stimulants.

Two thousand, five hundred years ago, Pythagoras looked at the horizon and knew it was, somehow, apart from the randomness of the immediate geography. The cult of geometry is now so deeply rooted it has become axiomatic. All I have done in these pages is to point out nature came first and in nature, there are better natural systems of organisation no matter how magical, or mystical, geometry may seem.

The early Greeks gave us the words "philosophy" (a love of wisdom) and "geometry" (to measure the Earth). They also turned their minds to the subject of politics. It is a great pity there are none around now to help with my correction of the parliamentary pyramid.

I lie on a warm towel, the Sun beats down upon my face, and I think about a cool ice cream. No one comes. Minutes pass and the temperature rises. My skin glistens with sweat. The Sun is frying my head. Where are they?

I sense a strange sight. Not far away, where the tree stump was, an unusual windbreak; a triangular cloth, quite large, three metres high and four metres wide. There is no name or message on it, so I decide to investigate. Who on earth would have a triangular windbreak, three by four by…five? No! It can't be … I peer around the hypotenuse and there sits an elderly gentleman wearing a woolly hat with side-flaps over the ears and a long, black cape. He looks up, calmly examining me with clear brown eyes. He has a long straight nose set above a white, woolly beard.

I bet he's hot!

When he speaks it is quiet, placid, unintelligible and Greek. I reach into the pocket of my swimming trunks for Galom's unitranslator, thus proving the Sun *is* frying my head and I have embarked upon another mental excursion.

"Pythagoras?" I venture.

The old man smiles and turns to look out towards the horizon. He says: "I am older than him."

"Socrates?"

The translator does not respond.

"Plato then."

The man smiles again, but shakes his head. He rotates his hand towards his chest. I must be close now. He gestures I should proceed a little further.

"Aristotle?"

Yes! He nods his head and his smile broadens, revealing a set of ivory white teeth.

Aristotle – I cannot believe my good fortune: Aristotle, the founder of systematic analysis, the first man to document over one hundred and fifty political constitutions.

"You are the very man I've been looking for," I say.

Another knowing smile: "So that is why I have been dreamed into existence. Sit down. I sense you are in need of guidance."

"Can you still guide me?" I enquire. "The institutions you studied do not exist today."

"Politics, my dear friend, is only one of my many interests. I have studied under and worked alongside the great Plato, a teacher of immense stature. I examined my fellow man and how he lives. To teach this to others one needs a framework; the subject must be divided, then divided again, and each sub-set solved and elucidated. I have combined this analytical technique with proper common sense to produce my discourse on the politics."

His credentials are impeccable, but are his ideas still of relevance today? I ask him.

"Our discussion," he replies, "may reveal this."

Where to begin? Imagine (as I am) that you are having a conversation with Aristotle, the most influential philosopher of all time; a man with a wealth of knowledge, insight and wisdom;

an expert of observation, analysis and argument; a real cool customer. Where does one start? Steady now. Focus.

"It's hot, isn't it?" Not a good start; just breaking the ice, so to speak.

Aristotle seems uninterested in such trivia. He asks: "What is the nature of your enquiry?"

Good; he is going to help me and guide our conversation. At last I begin in earnest.

"I am a humble electrician and ..."

"Ha! A fool," he interjects. "Citizens should not lead the life of hirelings. Such a life is ignoble and detrimental to virtue."

"I have to live, Aristotle," I protest. "Virtue doesn't pay the bills. I need money."

"Hand coinage is an artificial medium having no root in nature."

"Maybe so," I reply, "but it is very convenient, too. All manner of differing exchanges are regulated by its use. How else could people save?"

"Pursuit of monetary gain beyond the satisfaction of one's needs is unnatural."

Aristotle pauses, as if that is the end of this particular sub-set. Then he asks, "Does the present-day tradesman have any say in working affairs? Are you able to voice your opinions directly to your rulers? Are you involved in the decision-making process? Is the workplace democratic? Is there a forum to discuss your day to day activities?"

"We have safety committees and amenity committees ... oh, and unions," I say.

"Peripheral trivia," Aristotle replies tersely. "Yet a tradesman is still no better than a slave. It appears as if this has not changed. All this I stated two thousand years ago."

There is a long silence. Perhaps I will not like him; a lot of people don't. I decide to begin again.

"I've had this idea, about systems, conflict, choice and what to aim for in life."

"I understand," Aristotle continues to look out to sea, "you are looking for the way."

Yes, I suppose I am. I have never thought of it in those terms before. Aristotle turns to face me. "To answer we must decide what is the most desirable life, for if we do not know that, the best constitution is bound to elude us. I teach the well-being of all men depends upon two things: one is the right choice of target; the second is finding which actions lead to that end. Where in your treatise do you begin?"

It seems a long time ago, yet it had very humble origins. "I began my search for the goal of life, metaphorically speaking, driving around in a car park looking for the best spot."

Aristotle leans towards me and shakes the unitranslator.

"A chariot park?" he queries. "Strange."

"Do you believe a humble electrician could devise a novel system?"

"Pretty well all forms of organisation have now been discovered," Aristotle replies. "There are constitutions whose authors are sometimes statesman, or philosophers or laymen."

Laymen, eh? Perhaps I am in with a chance; but where is the best place to begin such a task?

"What is needed," says Aristotle "is a system which the people will be easily persuaded to accept, starting from the system they already have. Mind you, it is no less a task to correct a constitution than to create one from the start."

"Yes. I wholeheartedly agree."

Aristotle stares back out to sea. "You made mention of a goal."

"Yes." I am starting to like him again. "My definition was even all-round growth; no extremes permitted. I tried to visualise it like the expanding branches of a tree as it matures from a small ball of twiglets to a big spherical canopy."

The wily Greek ponders upon this for a while. "Very lucid," he says, as if he wished he had thought of my definition first. He looks skyward, peering directly into the Sun. "The expanding sphere is the most natural shape in the Universe. The body consists of parts, and all increase must be in proportion, so that the proper balance of the whole may remain intact, since otherwise the body becomes useless. A shipbuilder would not let the stern or any other part of his ship be out of proportion. Yet the end aim in any knowledge or skill is the greatest good."

"Sounds okay," I reply, unconvinced. "But it's not very specific." I provoke him for a change.

"In order to attain the good life," he continues seemingly unperturbed, "education and virtue have the most just claim of all. It is quite impossible while living the life of a hireling to occupy oneself as virtue demands."

Charming! He was perturbed after all.

"Look, we are digressing," I say, as if to call a truce. "What is the goal for the people, and for that matter, the constitution to aim at?"

Aristotle takes a deep breath ready to expound another treatise. "In my ethics I state virtue is a mean and the best life is the middle life. The best constitution must therefore be one which allows attainment of the mean to men of every kind; men whose education does not depend upon the luck of their natural ability, or of their resources."

"Is this your famous Golden Mean?" I ask.

"People will call it that. Yet, there must be some deviation from the mean, of course. If a political judgement needs to be made, answers will come from many quarters across the whole political spectrum, and on many occasions a crowd is a better judge than one informed man."

"Or woman," I add, but Aristotle is unlikely to know what I am intimating.

Aristotle shrugs his shoulders, then says it another way: "Each individual may, at times, be a worse judge than an expert, but collectively they will be better, or at least no worse."

Sounds reasonable: a few more referendums on major issues (like hanging) would be a good idea. Then instead of keeping mass murderers and sadistic child killers in better conditions than the country's senior citizens we could top 'em and remove their defective genes from the genetic pool. The truth is politicians are afraid to hear the "voice of the people" on many contentious issues. Only when that voice coincides with their particular dogmatic assertions do they incite it and act. When the collective voice differs from their ideology they ignore it and are allowed to because of a system where minority parties rule; representation is a sham and the people's expectations of the System are low.

Aristotle waits patiently while the above thought occupies my mind. I have invoked him, my pen commands him, but his thoughts are his own, rescued and preserved from the golden age of Greece.

"Aristotle," I attract his attention once more, "when you studied constitutions which did you feel was the best type?"

"There is so much to remember, my friend," he replies.

Silence; while he thinks of more ideas. He begins: "There were two classes of constitution. These were labelled oligarchic if they were too strict and master-like, and democratic if they were loose and relaxed." Today these would be similar to capitalism and socialism respectively.

"Which is best?" I ask eagerly.

"Patience! There is more to be said yet. You young ones scurry about so quickly in your search for the way that you overlook it many times before you discover it very close at hand."

Our eyes meet and we smile at one another. I nod submissively, acknowledging the truth within his words. His smile widens and the pearly white teeth are revealed once more. He continues: "Both capitalism and socialism are deviations from the best constitution. Those constitutions that are harmonious and well balanced are best; those that aim at the common good; those in accord with natural justice.

"Capitalism is the rule of the wealthy; socialism the rule of the poor. Everywhere the rich are few and the poor are many. Yet anyone who thinks in terms of absolute equality must also be wrong. For men are not equal in birth and wealth, courage or education. Superiority in birth and wealth ought to contribute to the quality of the individual. In fact, these qualities contribute nothing at all. The wealthy, superior in monetary terms, think themselves superior in other matters, too."

Aristotle seems to be rejecting both political wings. This means (since the answer cannot lie outside the boundary of the problem) perhaps the best constitution is nearer the middle. The middle way, harmony, balance; these attributes fill Aristotle's writings.

"I assume neither capitalism nor socialism fits the bill as the best construct. Then what does?"

I read in Aristotle's expression a measure of satisfaction, as if I have asked a good question at last.

"A constitution is really well made," he states, "if it looks like both and neither. It should be kept stable by means of itself and not through outside agencies. It is not doing that when only a majority of citizens want to continue the system, a condition which can be equally found in a bad constitution, but only when no section of the State would ever wish a different constitution."

This makes me think of the different branches of a tree all concerned with the tree's nourishment, no matter what direction they face. Compare this to the British duality, where capitalism and socialism have historically waged war for decades, tearing the country apart, stunting its growth. I ask Aristotle if the two factions will ever unite.

"Those incapable of existing without one another must unite," he says. Presumably if they do not, and the present see-saw political trends continue, the country will end up impoverished. Aristotle continues: "There is a common interest uniting master and slave. The unifying factor is correct education. Education which produces ideal people produces the ideal state."

"Why," I ask, "did you not compose a perfect constitution, as Sir Thomas More attempted in his book *Utopia*?"

Aristotle replies, "To set down the whole organisation of State in writing to the last detail would be quite impossible. The general principle must be stated in surety; the action taken depends upon the particular case."

Sounds like another piece of tree theory here: their growth is governed by rules, but the strategy is flexible. Thomas More's *Utopia*, with its built-in rigidity, never came to be.

"Nevertheless," I pursue my line of questioning, "you must have narrowed it down a bit. You must have a better idea than most of which is the best constitution; it was necessary for you to quantify in order to teach the subject of politics."

"Quite so," Aristotle replies, "but will your readers wish to hear all this? Wouldn't they prefer to be back on the beach with an ... erm ... ice cream? Perhaps they want to know how Linda is progressing; or read something more light-hearted and less tedious. Why should I impinge myself upon your book? What about a ..."

"No," I interject boldly. "This is no time for an intermission. You are here to provide moral support. Your views are akin to mine, yet if I pronounce them they settle with less weight. Extremists have already rejected the book; those unsure may want to hear us out. So continue. More details of your view of society, please."

Aristotle takes a deep breath. Our readers may wish to do so as well.

"You are in a determined mood, young man," Aristotle continues. "Eager to listen and perhaps learn; it is a pleasing combination to observe. Therefore I will proceed."

"In all states there are three sections."

"Nowadays, Aristotle," I rudely interrupt, "sociologists use a system with five socio-economic categories or classes."

"How quaint," is his terse reply. "As I was saying, there are three sections: the well-off, the poor and those in between. We have agreed moderation and a middle position is best. Likewise, to own a middling amount is best of all. This condition follows reason, and following reason is just what the exceedingly rich, handsome, strong and well-born consistently fail to do. Being unreasonable is also a failing of the extremely poor, the weak and those grossly deprived of honour. The rich are inclined to arrogance and crime on the large scale, and the poor are prone to wicked ways and petty crime. The rich neither wishes to submit to rule, nor understand how to do so, and this is ingrained in them from childhood. The poor, on the other hand, become too subservient and do not know how to rule. Both classes are therefore unsuitable to be rulers of a State. It is the middle citizens of the State who are the most secure. They neither covet the possessions of others, like the poor, nor do others covet their possessions, so they live with less risk, neither scheming, nor being schemed against.

"The best partnership in a State operates through the middle people. Those States where the middle element is large have every chance of a well-run constitution. For the addition of its weight to either side will turn the balance and prevent excess at the opposing extremes.

"The superiority of the middle constitution is clear also from the fact that it alone is free from factions."

All this makes sense to me. It is a welcome support of tree theory from a genius of a bygone age. But there are still questions to be asked.

"Why," I question, "if the middle constitution is so superior are most countries either capitalist or socialist and not a blend of the two?"

Aristotle ponders upon this. He knows the answer, of course, but he seems sad, as if he understands the answer too well.

"Where the middle element is large," he replies, "all is well. There arises the least division and factions among the citizens." He stops. That must have been the good news; now for the bad…

"However," (I told you so) "the middle element is frequently small. Whichever of the two extremes has control, those with possessions or the common people, abandon the middle and conduct the constitution according to their own notions. So the result is either democracy or oligarchy. A socialist or capitalist considers supremacy in the constitution as a prize, a victory to be won. Each side looks for its own advantage, not to the benefit of the State itself.

"So for these reasons, the middle constitution has never occurred anywhere, or only seldom and sporadically."

"Aristotle," I respond immediately, "let us assume a constitution favouring the middle emerges and it is good. How can we ensure it will endure?"

"Of all the safeguards that we hear spoken of as helping to maintain constitutional stability, the most important, but today universally neglected, is education; an education for the way of living belonging to the constitution in each case. It is useless to have the most beneficial laws, fully agreed upon by all who are members of a constitution, if the members are not trained and their habits formed in the spirit of the constitution. If not, just as one individual may be morally incapable, so a whole State will lapse into decline."

"This could be interpreted," I suggest, "as political interference in an individual's upbringing. Anything seen as indoctrination will not be tolerated. It will be regarded as a loss of freedom."

Aristotle replies immediately, "It ought not to be regarded as slavery to live according to the constitution, but rather as self-preservation. It is not by chance a State is made sound; that is a task for knowledge and deliberate choice. A State to be sound requires the citizens who share in the constitution to be sound. One needs to learn to be a citizen, just as a craftsman needs to be trained in his particular skill."

"Any attempt to imbue anything but the basic culture of our society," I announce, "will be met with profound opposition. You can teach children about Easter, Christmas and Guy Fawkes, but to teach them Thatcherism or Marx will not be tolerated. This would be seen as contrary to the concept of individual freedom of choice."

"It is not right," Aristotle resumes, "any citizens should think he belongs to himself alone: he must regard all citizens as belonging to the State. Each is a part of the State and must therefore bear a part of the responsibility for the whole. The education which produces an ideal person produces an ideal State."

"How do we produce the ideal person?" I enquire.

Aristotle looks down at the sand. His face falls; his voice becomes laden with gloom: "The depravity of mankind is an insatiable thing. For there are no natural limits to wants; and most people spend their lives trying to satisfy their desires." He sighs and this seems to raise his spirits slightly. "The task is not easy. Men become sound and good because of three things: nature, habit and reason. Man alone has reason and so needs all the three working concertedly. For men learn partly by habituation and partly by listening. After that it becomes the task of education."

Aristotle melts away ... I knew he was too warm!

Just then a little girl with ice cream around her mouth, and a smile from ear to ear, creeps up on her Dad and clasped in her hands is a bucket of cold, salty, seawater. Her mother stands watching, quietly chuckling while holding a rapidly thawing ice lolly.

Dad is still unconscious, dreaming, wishing he were a shade cooler.

Moments later his wish is granted.

Hindsight is an amazing thing. I do not mean thirty year hindsight; I refer to Aristotle's thoughts over two thousand years ago: "The rich are inclined to arrogance and crime on the large scale. The poor, on the other hand, become too subservient and do not know how to rule. Both classes are therefore unsuitable to be rulers of a state."

There is no doubt the Tories do not understand the struggles of the poor, as they have never been there. They have never lived it. Once you reach "poor" it is an uphill struggle all the way to escape. Help is necessary. It is unhelpful to have the rug pulled from under your feet, which always seems to happen at the worst time.

The majority of people in the country are amiable, educated, honest, working steadily, bringing up families, have some hobbies, keep their houses and gardens tidy, look after relatives and live a moral life. These are the golden people of society. But are their aspirations really represented by either of the two parties in the political system? I think not. I think they are divided and that is why others rule them and bleed them dry.

Meanwhile the British Constitution has no overriding goal as far as I can see. It is an open-ended on-going saga of ever evolving rules and regulations.

But changes can be made if only we can fathom out what is best to do.

Branch 16

Tree Counting

Holidays soon pass by.

Waiting at home is an overgrown garden, a house with its chimney in need of repair and a host of other summer-time jobs.

In work I am constructing a complex electronic panel with circuits to build and hundreds of wires to connect – preferably in the right place. This takes a lot more concentration than you might imagine and is a strain on the eyes and mind of this humble electrician.

At home Linda plays simple games now, like hiding her dummy and getting Tracy or me to find it. Then it is our turn to hide it and Linda screws up her eyes apparently tightly. "No peeping now!" We hide the dummy and say "ready", then Linda goes straight to the hiding place to retrieve it and we all laugh. Another game, which fascinates her, is taking objects, usually wooden blocks, out of one container and putting them into another. When the second container is full she reverses the process; good frisbee potential, I imagine.

Linda is not sleeping too well at night. We think she may be teething again. So around three o'clock each morning she whimpers and cries and only stops when she is in our bed in my place; bad habit: short-term peace leading to long-term exhaustion.

Meanwhile, in the World Cup, Brazil loses to Italy and England draw with Spain. Brazil and England are eliminated and I feel as if, once again, the cynical, dishonest sides overcome the honest endeavours of two proud nations.

All this stress and tiredness gives rise to a general dull mood. I feel subdued and phlegmatic.

June turns out to be the wettest one this century, but things improve as July approaches and gradually I rise out of the doldrums to tackle a few repair jobs on the house and achieve good progress on the project in work.

I still have some holidays to take and as a sunny week is forecast I go into work (on the Monday) and book the rest of the week off. And it is a glorious week. Local beaches and parks are visited, taking a picnic for lunch and returning home for tea.

By Friday Linda has caught a viral infection which causes her to lose her voice and cough vigorously. More sleepless nights lie ahead. One night Linda wakes coughing profusely. We bring her back into our bed, sit her up for a while and administer a drink of cool blackcurrant juice. She finished a bout of coughing, then has the audacity to tickle me! Unlike the adult, the young child doesn't understand she is ill and doesn't make a commotion about it. Not like her Dad! The mildest twinge and he lets the world know he's not very well.

So much for my main holidays. Now they are over, I slide back into a daily routine which reinforces a feeling of stability – no change. Nevertheless, there is still a great variety of jobs to tackle, leaving little time to consider theoretical matters. I can remember the summers of my youth: evenings playing tennis, or football, or (yes) even crown green bowling. No time for theory then either; too busy: result – sixteenth in class. In the winter: cold nights, dark nights, nowhere-to-go-nights; plenty of time to ponder over my homework: result – in the top five: first once, then Head Boy.

The thinking season is approaching. I have got to do a bit of tree counting before winter. Hence I make a determined effort, over a one-week period, to see how many branches a mature tree possesses.

Armed with a notepad and pen, I wander the lanes and venture into fields, circumnavigating the hedgerows and thinking to myself this is so pleasant an exercise does it really matter if it seems crazy at times? The therapeutic value of tree counting elevates it high above the clandestine world of politics.

The local landscape.
Nature has curves and fractal shapes, while man cuts hedges horizontally.

It is hard to imagine, when viewing the present day farmed landscapes, that twelve million years ago our planet was one homogenous mass of trees. Descendants of the lowly mosses and ferns, they ruled the Earth before our species emerged, helping to maintain life (as we know it) by exuding oxygen. Trees are the lungs of the Earth.

Trees are not as prolific now and to a casual bystander they may all seem pretty much the same. Today the tallest oak tree in the country has attained a height of one hundred and thirty-five feet, whilst the fattest has a waistband of over forty-two feet. Until iron tools were available, man found the oak an almost impossible tree to fell. However, one hundred years ago, oaks were a major source of building material.

In the eighteen hundreds there were ten thousand windmills in the country. Each required thirty oak logs in its construction. The main post of a windmill was usually a great baulk of oak, two feet, six inches square and eighteen feet high, weighing one and a half tons. The cross beams, to support the vertical post, were about one foot square and twenty feet long. By the nineteen hundreds mills were on the decline and one factor was a shortage of suitable timber.

The age of an oak tree was important to millwrights – the craftsmen who built the mighty windmills. At about sixty years old the wood is not hard or mature enough for heavy duty use. When one hundred and fifty years old the wood starts to become inflexible, but at one hundred years old the wood's properties are just right for a variety of uses; so oak trees have phases of growth, too.

I rest beneath a huge oak tree. The trunk is a yard in diameter and (I would guess) around three hundred years old. The foliage casts a large shadow on the pasture below, each leaf absorbing photons of energy until the mosaic is complete and sunlight is only allowed to dapple the grass when a breeze disturbs their placement.

Beneath the mighty crown one can look up through the main branches at the inner side of a verdant sphere which forms the food producing canopy of the tree. The history of the tree is contained in these branches: each twist and turn, the way it leans, each aspect marks an event in its life. This particular tree forks into two main branches, then within two feet divides again and so on. If it kept on doing this geometrically, a mathematician would consider this tree a perfect fractal. However, it doesn't and the result is a complex confusion of woodwork, a unique work of nature, an individual, which even computer graphics would struggle to depict unless fed by some semi-random algorithm.

I continue my counting, wading knee deep in tall grass. The lush hedgerows are mainly horizontal hawthorn and holly, but occasionally wild rose and pale yellow honeysuckles add a splash of colour. Common ivy intertwines in the hedgerow and occasionally ivies climb the trunks of the trees, thick stemmed and clinging high into the canopies. I don't think this is a good thing for the trees. Surely they must drain their resources?

It is not all peace and tranquillity strolling along the lanes. Flies buzz annoyingly around my face and at one point I disturb a dozing farm dog on a nearby track and he barks and barks until I retreat across the field.

I manage to examine one hundred oak trees, selecting the sample size purposefully so the results can readily translate into percentage terms. A histogram of the result is shown below, though it will need some explanation.

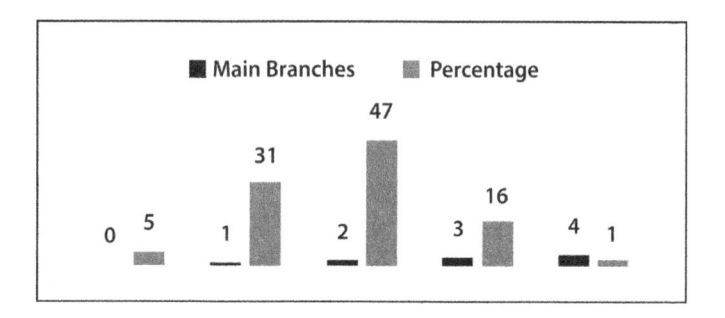

Main Branches	Percentage
0	5
1	31
2	47
3	16
4	1

Five of the sample were simply ovoid in shape, or spherical enough to be classed as maiden trees with *no* main branching and just a single trunk with all other smaller branches radiating from this tapering stem.

Thirty-one trees were so badly shaped and warped they could not be placed in the maiden category. Yet neither did they have a major division of the trunk. On occasions some were so badly twisted their single trunks were almost horizontal. I faced a dilemma in my categorisation. Fortunately I knew what to do about it (it's the kind of thing politicians do daily) I ignored the facts and made my solution look neat, this being an unwritten rule of statistics. I therefore decided I would class the badly twisted trunks as "trees with one main branch".

Forty-seven trees had a major division into two main branches. This is not surprising. Any decent gardener will tell you if you pick out the terminal shoot, two side shoots appear from the leaf joint below.

Sixteen trees had three main branches and one large tree trunk divided into four main branches.

What does all this mean? Quite frankly, very little indeed. Forty-seven trees had two main branches and forty-eight didn't, but the most common number of main branches in a mature oak tree structure was two. As you will recall, it is reputed having two main parties is the basis of a strong parliamentary system.

"Aha!" I hear you establishment buffs exclaim. "Case proven. The existing system is a good natural construction after all."

Perhaps beneath the duality I have dealt with, the conflict between public and private concerns, staff and industrials, left and right wing, there is (after all) a natural underlying force. Are dualities and conflict unalterable frames of reference for Earthlings? In a lot of cases this appears to be true, as if conflict itself is the driving force behind progress. However, it would be hard to discover any cohesive progress arising out of the political diversity which has held sway since the last war. Yes, there has been material progress, more for some than others, but social progress has stood still with the two sides entrenched in their class ruts. Really, the whole of the Western world is philosophically bankrupt. Instead of searching for points of compatibility, new trenches are being dug.

A healthy tree can have two main branches and still flourish in harmony and balance. Yet we have had a two-party system which has accelerated our decline in international matters. But both main arms of a real tree, although disagreeing on the path to take, have one overriding function which is to improve the tree. Our two main parties only foresee a good future when one wing is destroyed. Each party can find nothing good in the other. Neither looks over the pyramid's apex to see the benefits on the other side. And although two main branches are common, those trees with three or four can still support healthy canopy systems. Two parties are not, as we have become accustomed and *conditioned* to believe, the definitive state of political affairs.

Our encounter with Aristotle was a lengthy one. This branch of thought is purposely short while the reader recovers. But before I close I want to recap on the main attributes of pyramid hierarchies and tree systems. See if you can decide which system Aristotle and I favour.

	Pyramid Hierarchy	Tree System
1	Abstact physically rare	Natural organic structure - prolific
2	Built with discrete blocks	Continuity between various elements
3	Rigid rules	Strategy flexible within limits
4	Shallow foundations	Deep roots - heritage
5	Information is one way - down the line	Fed from both ends with fluidity
6	No definable good shape	A known good shape; generally speaking, round
7	Stability assumed	Evolved for change
8	Only one aim. Usually excess profit	Goal - all-round balanced growth. No extremities
9	Little real moral or social conscience	Tries to exist in equalibrium with environment
10	Perpetually forces growth	Succumbs to natural cycles - growth and consolidation

An explanation regarding "I go into work (on the Monday) and book the rest of the week off" – In those days, with large teams and strength-in-depth this was accepted practice. Later on, when the workforce was "lean and mean" a sunny day off was frowned upon. Indeed, any time off at all caused stress to those left behind and to oneself on return to a bulging in-tray. The work life balance was in bits.

If you think it's all about hard work, you need help. It's being effective that counts.

Now I am wondering if there is some pattern to the thirty year cycle that seems to repeat throughout this work. Not only in terms of recession, but in extremes of weather like the recent blizzards and deluge. Last year, after an initial drought, the country was flooded all summer long. The country wide water table was the highest on record.

It was interesting for me to note (on reviewing this chapter), that even thirty years ago, I believed material progress had outstripped social progress. This means, in total, I feel the organisation of British society has made little positive advance for *sixty*, not thirty years; or, in other words, during my lifetime, i.e. since the nineteen-fifties.

Moving on, and this will seem obvious when I point it out, no two trees are the same. They are, like human beings, true individuals, shaped and formed partly by their immersion in life and the winds of change. We are all unique.

Don't let too many outsiders cling to your tree of life. They will drain your resources.

Branch 17

Loose Ends

Events move on rapidly. Linda start attending playschool, British forces recapture the Falkland Islands, a general election is held and the coal miners go on strike.

A baby brother arrives for Linda to play with, triggering six months of physical and emotional strain. Linda has been the focus of attention all of her life, but now finds herself an equal first, somewhat of a demotion perhaps. The young lad, Owen, needs a lot of attention and he gets it. Linda tries hard to help, but when the pressure is continually applied our patience threshold is low. Fetching things is still a game to Linda; to us it is a chore. Any delays irritate: "Just bring it here." And I feel I lose my temper more than I should.

Young Owen is another branch of the family tree and for a while the balance is disturbed.

Linda brings all manner of ailments home from playschool to batter our weakened immune systems: mysterious rashes our doctor is unsure about: coughs and colds and a virus or two to purge the system. This promotes the unhealthiest period of our lives, which saps the last of our physical and mental energy. I have sinusitis, continuous colds and catarrh from December through to May. Yet, for reasons we will soon encounter, I struggle on and, against all the odds, do not lose one day of work through illness or debility.

Meanwhile, in work big changes have taken place. All redundancies were filled with volunteers. Now there is a *shortage* of labour. Management announces (unilaterally and without even the civility of consultation with the employees or their representatives) that contract labour will come onsite and tackle priority work. The workforce is incensed and starts a work-to-rule: this is the pure application of pyramid hierarchy rules of the System to the System. Management seem dismayed with delayed projects, which is always the case when the workers abide strictly to management's rigid rules. Normally the workforce is only guided by rules; they operate and function on spirit and goodwill, which is obviously absent when industrial action emerges.

Last winter, when the weather was really bad, workers were marooned by snowdrifts at two power stations in Kent. How did they respond? I quote from a newspaper report: "There was a fantastic response. Everyone mucked in and the rule book went out of the window. People just did what was needed to keep the place running for the consumers' sake." The item continues: "Conditions inside the station were grim. Pipes froze. Men got what sleep they could on office floors." A senior employee then comments: "I've worked for the Board for twenty-five years and all the old spirit came right back. Guys worked all night to keep internal roads clear. It was phenomenal. They were all shattered." Sounds like a good commendation for a small portion of humanity in an extreme situation.

During the present Conservative-enhanced recession an awful lot of spirit has been dissipated. People have been retreating back into their boxes to shelter behind the rules and (hopefully) justify and preserve their jobs. This effect, as I understand it, is exactly the inverse of the Tory objective.

Meanwhile, back in the factory after industrial pressure, an agreement is reached whereby the recently sacked apprentices are sent letters inviting them to return on an annual contract basis. Many do – there is nothing else.

Surprisingly, while others see this as the beginning of the end, I consider it to be a new beginning from where we may see a possible upward swing in a new cycle. Consequently, I feel this may be a good time to try to foster a bit of new team spirit and I make a concerted effort to interest management in a "canopy committee" style meeting. Of course I do not mention it by this name, since management have not read this book and by choice would never want to. It is soon made plain both management *and* the unions are opposed to my idea and I realise this is a sure indication the idea has *considerable* merit. Union enthusiasts are worried meetings where their members are allowed to speak for themselves might undermine their authority. Management are full of the Falkland spirit and are acting resolutely, even if resolutely wrong. Co-operation is "out"; conviction is the new "in" thing. So after a six-month word-in-the-right-ear campaign I concede defeat. It is nothing to worry about. The climate just isn't right for any form of consensus. It is not my loss.

I asked earlier in this work: "Who would grumble if everyone was kept up to date and on their toes?" In the main the answer seems to be middle management. This is one in the eye for Aristotle, since he might consider middle management to be a mean between extremes and therefore the best on offer. British management have traditionally shunned any socially progressive ideas. They don't see it as part of their job to innovate. Someone else gets paid to do that. It is a mystery why this should be. Japan has held "canopy" style meetings for decades. Japanese companies in Britain pursue this philosophy and they thrive. British workers love the idea and readily take to them, too, but British management do not.

I find it a bit of an irony information meetings cannot be sponsored. For if you wished to start a squash ladder, a crib club, a chess section, or a computer club, all you need to do is pin up a notice asking for volunteers. But try to promote a meeting to discuss aspects of your employment and management starts to worry about your ulterior motives. They have not tried asking for volunteers to sample opinion – this method isn't mentioned in works' instructions. No, they just won't bite. The pool of progress remains still. No surface ripple: no waves; just as they like it. Onlookers may sense an odour of stagnation.

As part of the factory's reorganisation, my section is cut from ten members to three and it soon becomes apparent we are now so efficient we can no longer cope with the workload. The section is termed a "section under stress." All spirit bleeds away and our work becomes a holding operation against a seemingly endless tide of priorities. On one occasion I am taken off priority work, switched to an urgent job, then moved again to attend to a safety hazard. Well, it's nice to be wanted.

Running consecutively with these events, my best mate, colleague and prodigy, young Keith, nears the end of his fully-trained phase. He is ready to leave. I can sense it; he knows it, but may not realise why in terms of the tree-system's theory of phase development. A person from a lower phase of development has great difficulty in understanding someone from a higher phase. This fact is more popularly known as the generation gap.

The decline in working atmosphere, the prevailing impersonal attitude and the new style of working – switching from job to unfinished job – finally settles matters. And so he gets another job and departs: a tragic loss.

There is a position, somewhere along an imaginary line, where efficiency and quality collide, or at least the two do not mix. At some point high efficiency gives lower quality for someone.

Now full overtime returns; ironic, isn't it? The overtime is the reason I do not lose a day's work through illness, as it provides a welcome respite in my personal economic recession. We can but hope for a happy ending to this branch of the story.

There are other branches that need further attention, too.

The political branch and the Parliamentary tree were left in a state of limbo while I gathered moral support from Aristotle. His rule of the Golden Mean seeks moderation in all things. Although this idea is not flawless and is unappealing to the fervent idealist, it does have considerable merit. Aristotle believed the answer to most conflicts centres on the mean between opposing extremes. On a scale of nought to ten this does not mean the answer will always be five (if you see what *I* mean). The answer may be anywhere from, say, three to seven, but will definitely be somewhere between the two extreme points of view; the answer rarely lies outside the problem.

Now (in this country as I write) the two main parties are as far apart as I've known and the gap is widening. What happens next is quite stimulating. A new central branch emerges to fill the gaping void left by the two diverging parties. I wonder if the resulting conflict between three parties will be a bloody despicable affair, mauling one party into submission, returning us to the old two-party system and back to square one.

The new central branch is named the Social Democratic Party – SDP for short. Soon an alliance is formed with the Liberal party, which is a different branch also occupying a central stance. Initially they receive a lot of support, but this wanes after the Falkland conflict.

This is not a politically inspired book; it is about framing a philosophy to try to help one through the turmoil that engulfs our lives. Inevitably, politics must emerge at some point in the discussion. Different philosophies underlie, or lead to, differing political viewpoints, as we have seen. Socialism is Labour's ideal, while Conservatives pursue the capitalist goal of the unfettered market.

The SDP is the nearest thing yet to a political interpretation of tree theory illuminated in these pages. They set themselves twelve tasks to aim for, a hard core of rational pragmatic ideas and policies, blending the best virtues of the two main parties with some original thoughts of their own. They are the only party free to harness ideas from across the political spectrum, their funding being free from union or business obligations. Their membership try to pay their own way, while voting for the leadership is not done by block voting or by secretive committee, but by each individual member in the comfort of their own home by postal ballot. In many ways this makes the SDP almost above politics, set apart from the present cartel of the big two. The SDP's formation drew upon its members' shared belief that eliminating the class antagonism (the two main parties foster) is the key to national success. In tree terms, they wish to reinforce a spirit of national unity where the whole tree is considered prior to the individual main branches. A bit less I'm-all-right-Jack and a bit more altruism. What they lack, in my opinion, is a recognisable supportive philosophy and, with respect, I feel they must find this before they can progress. I also think their ideal philosophy is waiting to be uncovered among the modern concepts of system theory.

On paper the economy seemed to be recovering. Frankly, when you have been driven to rock bottom there is only one way to go. The recession (we are constantly told) is over and the official Opposition (Labour) are now in disarray, an ideal time for the incumbent government (Conservative) to call a general election.

The media merchants get to work on the two vital sections of the electorate, the floating voter and the apathetic abstainer. The former comprise fifteen per cent and the latter twenty per cent of the electorate. Factors which decide who shall govern can be mere innuendos perpetrated by partisan advertising agencies.

The Conservatives are returned to power. Though fewer supported them this time they increased their number of seats in Parliament! This is not because they are well liked, but simply because the socialist vote is now split. Thirteen million voted Conservative, eight million voted Labour and seven million supported the SDP/Liberal Alliance; so thirteen million people are for the government and sixteen million against it and another minority government is foisted on to the country because of the crazy first-past-the-post election procedure. Another minority government imbued with the power to rule absolutely over a majority who have rejected their monetarist philosophy.

Although the SDP/Liberal Alliance polled seven million votes they return only twenty-three Members of Parliament. Put another way, each Conservative MP needed 32,000 votes to get elected, each Labour MP 40,000 votes and each Alliance MP a staggering 338,000 votes. This is a bad system of enormous proportions.

	Labour	SDP/Liberal Alliance	Conservative
MP's>	209 seats	23 seats	418 seats
Voters>	8,000,000	7,000,000	13,000,000

Polarisation politics is a symptom of pyramid hierarchies. The House of Commons debating chamber has two sides; enemies can then face each other in opposition. No provision is made for more than two points of view. The modern debating chambers are nearly always crescent shaped, from the United Nations to the European Parliament. But we in Britain are stuck with the House of Commons: a cathedral of conflict not suited to consensus. Perhaps Guy Fawkes knew more about architecture than he has ever been given credit for?

What does the Alliance think of this situation? They think the system needs changing. Indeed, they are the only party seeking change. The two main parties are quite happy with the status quo, where they can take it in turns to cock up the country. However, the Alliance also knows they have to work within the old system and beat it before they can talk of change.

What does the electrician think? The electrician knows the electrical resistance of a tree decreases non-uniformly from top twigs to the base of the trunk. Granted, not a lot of people know this fact. Permit me to clarify my statement. At the top of a tree, at a height of (say) forty feet, the resistance to earth is 400,000 ohms. Lower down, at twenty feet, where the trunk diverges, the resistance is now in the order of only 10,000 ohms. The tree's resistance does not vary linearly with height and this may be expected if you consider the large difference in cross sectional area between twig and the trunk.

Put another way, when a tree trunk divides the cross sectional area of the supporting bole equals the sum of the cross sectional area of the emanating branches above. This sounds a bit Pythagorean. An illustration may help...

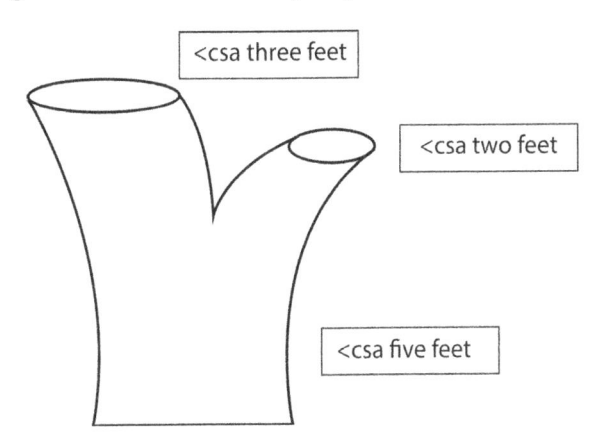

```
          <csa three feet
              <csa two feet
         <csa five feet
```

This same simple rule applies all the way up the tree. Each division is, therefore, nourished in proportion to its needs; clever eh? This makes the tree quite an efficient system.

Now then, the Parliamentary tree is, as we have seen, a product of a poorer, less efficient, unrepresentative, divisive, first-past-the-post system. It can be readily seen the middle branch is not receiving the support it needs, deserves or justifies. Seven million voters are represented by only twenty-three Members of Parliament. Bad system.

The present MP's priority undoubtedly places the wellbeing of his party before those constituents he represents. So much so that his individual influence on an election result has been assessed at a mere five per cent. In Parliament he votes as his party dictates, not as the people he represents may wish.

The party is the main symbol of its class: Labour, the vanguard of the working class; Conservative, the ally of the business class. These two are locked together in a wrestling match which debilitates the country as a whole. The miners' strike was a class war between these two old adversaries. Yet the two classes will always and forever be inescapable parts of the overriding populace: there will always be those who "think" to earn a living and those who translate those thoughts into action. These two elements are present in every society yet devised.

With our present system, where party dominates the MP, the class war is reinforced. The Government branch, in particular, wags the people, while the other parties are under-represented and unable to wag anything.

In a real tree the leaves supply the food and are nourished in a two-way exchange in proportion to their needs; a proportional system of representation, which is a system of voting used throughout Europe. It is generally agreed proportional representation is better adapted than any other voting method to produce an image (in the Chamber) of the feelings of the nation as a whole. And this is the aim that any good electoral system should achieve. First-past-the-post fails on a number of points.

In 1951 Conservatives polled fewer votes than Labour yet had a majority over all other parties: three hundred and twenty-one to three hundred and four seats. In 1974 Liberals polled twenty per cent of the vote and won only fourteen seats, less than three per cent of the seat in Parliament. In one by-election the following votes were cast: Conservative six thousand, Liberal five thousand-five hundred, Labour five thousand and Scottish Nationalist two thousand-two hundred. The Conservatives thus gained the seat although twelve thousand (sixty-six per cent of the electorate) did not support them. In this situation the winning MP is termed a minority member. That is, although he was first past the post, the majority did not vote for him.

In the 1964 election there were two hundred and thirty-two minority members returned to Parliament – about one-third. Finally, in 1968 only forty per cent of the general public knew the name of their MP.

First-past-the-post exaggerates the power of the largest party and diminishes that of the smaller ones. Proportional representation is a unifying force, where shades of political opinion are truly reflected in the debating chamber. This system eliminates extremes, restores balance and brings under control the present divisiveness which is tearing the country in half: North versus South, unemployed versus employed, working class versus rich. These are deep-seated problems which need to be responded to with a philosophy outside the confines of the polarised political debate. We have to revive the concept of fellowship, community spirit and consensus; the spirit born in us, then educated out.

An example of this spirit appeared in the playschool potato and spoon race. Potatoes are used instead of eggs for obvious reasons. Linda and her friend Samantha are on the starting line, potatoes on spoons; giggling. Ready, set, go! They are off and oh! Linda's potato has fallen already. Now what? Samantha has stopped, too. She picks up Linda's potato and between them they manage to get it back on to the spoon, but then Samantha's potato falls and Linda helps to pick that up. By this time the race is already won and the two girls cross the line much later, still giggling. Will they exhibit this altruism after a few years of schooling? Or will each want to be first past the post?

Millions of years before proportional representation was conceived as a means of voting, real trees were using proportional systems to consolidate their designs. The present Parliamentary tree must accept a more representative design before it too can progress and mirror the aspirations of the citizens.

The existing two-party cartel detracts from the overall long-term health of the nation in pursuit of biased short-term gains in every important field of activity.

In these pages I have painted the pyramid as an evil hierarchy; one that is impersonal; one that stifles creativeness by limiting the diversity which is within us all; a structure which clones people to perform irrational tasks to fulfil a one-line-rule for some senseless purpose. Yet all this is not strictly true. It was, however, necessary.

The difference between pyramid hierarchies and tree systems is not precisely one of evil versus good. It is a choice between a moderately good system (the pyramid) and a far better one (the tree). Pyramid hierarchies have provided the stability necessary for social organisation to be nurtured during its foetal stage. Now we are entering an age where the social traumas of the twentieth century cannot be contained by the rigid hierarchies of a bygone age. A new order must evolve. The turning point has come. It is here with us now; yet powerful Establishment forces are trying to conserve the past. The status quo must be swept aside, smoothly and purposefully. And it *will* be *only* when this new wave is properly understood by the masses. Not as a tidal wave of destruction and revolution, but as an older, friendlier swell that swept our ancestors out of the oceans eons ago and is now set to return to refresh our souls.

You'll be glad to read that I'm putting all this "politicking" to bed soon and moving on to far more interesting things. But first here is an update on representation.

Quite recently, in 2013, the people of Great Britain had the chance to vote for a more proportional form of representation called the Alternative Vote, or AV. It was never properly explained and overwhelmingly rejected.

Now we have a coalition government, but this is not anything to do with proper representation or the blending of disparate ideas. It is an amalgam of non-winners doing things (policies) not even in, or totally contrary to, their individual manifestos. You could not act this way in any other form of organisation except in a cobbled together government. If you did, it would be illegal.

Democracy is failing at every level, here and throughout the world. As mentioned before, even the United States of America is split, too, politically paralysed and divided amongst itself, with a small faction of Republicans holding the bi-partisan, democratic process to ransom.

And yet I read first-past-the-post is the best form of electoral system! Not because it is beneficial, but because it is the *best* system for removing bad government. This must be due to the fact that small swings of opinion have disproportionate results.

But, surely, we want to vote within a system for something we believe in; not vote merely to keep out a party that we don't believe in. That is too negative. That is how we have ended up in this mess.

We have to get away from indirect democracy and first-past-the-post back to direct representative democracy and the signs are this may evolve out of social networks into the mainstream. Polarisation has to be eliminated. The age of the tribal party should come to an end and be replaced with more free voting by responsible people representative of the community as a whole. If half were female, it would be a step in the right direction.

New Zealand voted for electoral reform in 1993. The populace was dissatisfied with both main parties and chose Mixed Member Proportional Representation

(MMPR). As a result, the NZ Parliament became "more colourful".

Within MMPR you have two votes: one for an MP of your choice and the second for a party of your choice. It's not that complicated, yet the country still struggles to understand the system. In a referendum in 2011 MMPR was resoundingly retained by the New Zealanders.

Germany uses the MMPR system too and I don't see them struggling to get by.

So how shall I advise you to vote in future British first-past-the-post elections? Capitalist or socialist; right or left; heads or tails? Answer: it doesn't really matter. Each party will play out its alternative existence on either side of the same coin while the people in the middle are pulled apart. It's the coin which needs changing. Until that's achieved, social decline and division will continue against a backcloth of apparent materialistic gain.

Now to far more interesting matters ...

Branch 18

The Way

Superficially, man is a very clever animal. I used to think I was clever. I mean, I knew $E=MC^2$ and all that square of the hypotenuse stuff as if it were second nature, then a little four-year-old girl started asking me questions. Simple questions: so simple adults never think of them and if they did would not give them a second thought anyway.

"Dad."

I'm eating my tea, have a mouthful of shepherd's pie and cannot answer immediately.

"Dad."

Linda will go on saying "Dad" until I respond positively. Time seems to travel faster for her and she can usually insert four or five "Dads" before one of my swallows.

"Mmm?" I frown ... well, go on then.

Linda has stopped eating and is staring at her fork. She is in a phase (way down her young tree) where the novelty of eating has worn off.

I swallow.

"Yes, Linda?" I wait. "What?"

"Why ..." she begins and stops.

Yes, very good. Should I chance another mouthful of pie? I do.

"Dad ... Dad."

Aw gawd.

"Mmn-hmm?" (What?)

"Why do forks have slots?"

Why do forks have slots! Is that it? Is that all!

"Well because ... " It's obvious, isn't it? ... "Because if they didn't have slots ... " Yes, that's a good method of approach,

"… then …" (Come on dear reader, you know $E=MC^2$; what about this one? "…it would be a spoon instead." Satisfied? Linda seems to be and that is all that matters. But before she can do any more thinking I add: "Now get some pie on your spoon-with-slots and eat it."

Some while later, the exact chronology evades me now, Linda and Owen are having a communal bath. Each is playing with a plastic bowl, filling them with water and soap suds and making pretend soapy shepherd's pies. Owen uses a toy plastic boat as a spoon. Linda tells me to get a wooden spoon for her pie.

"Get a spoon?" I intonate. Bossy Miss Linda has omitted a little word.

"Please," she says and off I trot to the kitchen to find a large wooden spoon.

There are no sharp edges to the spoon and it will fit her hand comfortably. She starts to make another soapy pie, and then another question fills her mind.

"Dad … Dad."

"Yes." I'm getting better, only two "Dads" before I open my mouth.

"Why does the spoon go dark in the water?" she asks; another simple question.

I peer over. "Whereby?"

"Look."

Sure enough, the dry part of the spoon is a lighter shade of wood than the wet half. Of course it is. Everyone knows wood goes darker when it is wet, but why? I hazard a guess: perhaps when it's wet it absorbs more light.

"The refractive index of wood is less when it is wet," I pronounce.

Linda continues playing, happy an answer exists, even though she doesn't understand it and, frankly, I barely understand it myself.

There are more questions to come. The majority (naturally) are readily explainable in terms she can understand, but sometimes she hits on a big one, a puzzle even Dad needs to think about.

"Come on, Linda, it's getting late. Time is moving along."

"Dad ... Dad ..."

"What?"

"Why doesn't time stop?"

Beats me. It has never stopped while I have been down here. Some days do seem to go slower than others. Those days when there is nothing to do because of bad weather, or, in work, the stores has run out of cable, or you are waiting ... waiting to fulfil an appointment. Time does alter as the speed of light is approached, but Linda would not be impressed by such an answer and may ask further questions leading me into a black hole of my own making. I settle for a simpler solution.

"If time stopped we wouldn't have birthday parties or birthday presents." There! She can understand that. A good answer?

I carry on tidying-up and pick up a few more plastic bricks. It has gone awfully quiet – not a good sign on occasions.

"Dad ..."

"Yes," I reply quickly.

"Dad, how old are you?"

"Thirty-five."

"Why?"

I frown: "Well, that's how long I've been here."

"In this house?"

"No. Well ... not really. On Earth, I mean." Now I'm getting confused. "That's how long I've been here on the Earth. Thirty-five years. See?"

There is a short silence again while this sinks in. Then Linda asks innocently: "Why, Dad?"

"No one knows," I tell her and she accepts this as she can recognise from my vocal delivery and intonation I'm telling the truth. Children can detect these things.

She goes to bed. Owen is already fast asleep, leaving me sitting there holding a Lego piece whilst slowly falling into the depths of a black hole.

It's just not good enough; thirty-five years old and still no idea what I am doing down here. Of course, due to tree theory, we can now decide if the kind of life we are leading is aiming in the right direction: are our main areas of activity balanced; are we expanding too fast, or too slowly, or not at all? Is this dual goal of balanced growth attainable if our present situation continues? Are we trying our best in each branch of our activities?

Now we have some guidelines, some frames of reference to ponder upon and check our development, we are still faced with a problem. Are the guidelines in keeping with the purpose and what is the purpose? For Linda has just asked me, in a tangential sort of way, what do you believe in?

Tree theory now stumbles upon a major problem of communication; for if I reply I believe in trees my sanity would be questioned. If I elaborate and instead say open-ended, multileveled, dynamically balanced systems of finite duration ... my sanity would be questioned. Not good enough.

The void Linda has unwittingly exposed is filled, for the masses, by religion.

There is a variety of differing beliefs, but I generalise and split them into two main geographical categories: East and West, with the dividing line roughly around Pakistan. There are differing sects on each side and each has an estimated number of followers in millions as on the following page:

Predominantly Western beliefs		Predominantly Eastern culture	
	Followers		Followers
Sect	Millions	Sect	Millions
Jews	15	Taoists	30
Eastern Orthodox	90	Shintoists	60
Protestants	325	Buddhists	200
Muslims	500	Confucians	300
Roman Catholics	585	Hindus	500
Christians	1,000		

The sects on the left believe in God and are all descended from the writings of the Old Testament. Those on the right do not believe in an omnipotent god, though some have many deities, while others have none.

You may be surprised to find so many people do not believe in God; an estimated one thousand, million people (in the right-hand column) do not worship Him. Indeed, in China there was no ideograph, no word that may be fairly translated as "God". The term is strange to them, even though they are residents of the same little planet (unlike the imaginary Galum) supposedly created by Him. Why is there this geographic separation of faiths? No one knows for certain. I feel the stimulus which inspired the Semitic tribes of the Middle East never reached eastern Asia. The Asians instead had to work out philosophies based upon observation rather than faith and the hand of an omnipotent God is not readily visible in nature. By this I mean nature's cruel indifference to who lives or dies; floods, droughts, diseases, all these phenomena can also be seen as the manifestation of a neutral impassive force rather than the purposeful acts of a "merciful" God.

Since no proof exists in either case, we can only assess our position subjectively and individually, but we may not be given a chance to do this. For the same education which taught us $E=MC^2$ has also taught us (in the West) a Christian theology to the exclusion of everything else. Tracy and I have already been approached by our village vicar to see if Linda will attend Sunday school when she is five years old. I was caught in the net when I was young and went to Sunday school and now I think Jesus taught very good ethical and moral codes which I largely live by and accept. Why then have I no faith in God?

The decline in my unquestioning belief started with the realisation the bad do just as well as the good – sometimes better – and get away with it more often than not. Then, later on, it became apparent the one God was understood differently by a multitude of factions. Some envisage a divine trinity in Heaven, others venerate the Virgin Mary. To many, the one God is Allah. Some believe in an-eye-for-an-eye, while others want to turn-the-other-cheek. The Jews are the only sect remaining who believe in a monism – the Lord God, Yahweh – without any other middle men, prophets or virgins involved. How could God permit such diversity?

How could this merciful God allow millions to lead a life of starvation and disease? Why won't He intervene?

Perhaps God, as defined by Earthly theologians, does not exist. Perhaps the Churches which preach no birth control in an overpopulated world, abhor Apartheid yet invest some of their vast wealth in white South African institutions, have got it all wrong; after all they have been wrong before – remember Galileo?

They have apparently been bequeathed the greatest message the world has ever received, yet they seem unable to dispense it convincingly or in a contemporary way. They have created religion not to free the spirit but to imprison it.

How can a humble electrician have faith in the Word amidst a sea of variables? How can he build up respect for a God who is manifestly impartial to his creation? He cannot. I am looking for something simpler, yet deeper perhaps; a creed untarnished by denominational disputes; a fundamental way in keeping with the fundamental law of the Universe.

I do not feel we can dismiss God with certainty, but I do disregard the derivative Western theology and think it is divisive, hypocritical, cant, complacent and doctrinaire. The message has been lost in the turmoil of divine words; the average man can no more understand it than the working class can interpret *Das Capital*.

What about the religions of south-east Asia? Is there anything worthwhile out there? A relative profundity of deities (saints), gods, spirits and demons are worshipped. Dead ancestors are venerated, too, since all Eastern religions believe in cycles of birth, death and rebirth. Hindus practise a caste system, a kind of religious class system, where movement from one caste to another occurs through death and rebirth. Progression, or regression, through the castes is dependent upon one's Earthly activities. Good actions favour one's progress; poor behaviour lessens one's prospects.

There appears nothing on the surface to whet the tree theorist's appetite; just a baffling array of gods. I get the impression different sects used to show slightly more tolerance for each other than in the West, but I am unqualified to judge on that matter.

It is interesting to see (in over-viewing these ideas) the simultaneous emergence of religious concepts in separated regions during various epochs, as if religious events have occurred in phases: first drawings and paintings to capture the souls of the hunted animals followed by ritual sacrifice and ceremonies: then a belief in an all-powerful concept. In the Middle East it was God; in China it was Tao.

Tao is called The Way, or The Principle. Tao is the underlying force of the Universe, if you prefer, the *ultimate* system. Those who work in accord with the Tao are enriched, while those who oppose or ignore the Tao are diminished. How did this philosophy arise?

The setting is a lush, green valley surrounded by high snow-capped peaks set against a backcloth of clear blue sky. Clouds and mist form around the periphery, then condense, giving birth to steams and rivulets which cascade down mountain slopes, bubbling over smooth pebbles, crashing against larger boulders, flowing ever downwards into the valley basin. The pure water nourishes rich grassland and trees.

Here, surrounded by and infused in nature, the sage lived: the Wise One: the Old Master. His house was a round makeshift hut, walls of thorny bushes and a roof on which grass was growing. A mat, fixed to the pliable branch of a mulberry tree, served as a door. Two round jars, with their bottoms missing, are set into the wall and covered by clear stretched cloth to form the windows of the hut's two cells. The roof leaked and the floor was damp.

The sage understood The Way; the nature of things and all that was around him. Travellers said he could speak to animals. I prefer to think he understood their habits so well it merely seemed like that. The sage did not appear to do a lot, being content to sit and think and observe. He was so wise, due to this inner contemplation, princes and kings offered him positions of high ranking in their courts, but the sage preferred to keep a low profile and lead a more humble life.

The result of all this contemplation and study was a realisation nothing ever remained the same. Everything is transitory; all phenomena are subject to change. This seemingly obvious insight led him to think about how changes occurred.

There were three states of change: first, non-change. This does not mean standstill, however, and is not the same as lateral drift. As in tree theory, there is always movement; nothing is absolutely at rest. Rest is only seen as an intermediate state of movement, or latent movement; a relative base to measure change by. Like the stars in the night sky, seemingly fixed, yet understood to be in motion.

Second, there is cyclic change. This was most important to the sage: the cycles of nature, day and night, lunar cycles, seasonal cycles. From these natural events the sage extrapolated many other ideas in differing areas. For example, movement precedes rest, rest precedes movement; another cyclic change.

Life and death were also considered to be mere changes in state as part of another on-going cycle. Prior to birth there is non-existence, out of which existence arises. This is followed, once again, by non-existence.

Third, there was progressive change; a steady non-cyclic change. The gradual erosion of a mountain peak is an example here. There is a progressive change within each of us as we age. We develop, consolidate in a certain direction, grow more rigid and then decline. Although this progress is now measured against linear time, dependent upon how often some caesium atoms vibrate in an atomic clock, I do not consider our development to be linear. Indeed, it most certainly is not. A child learns at a fantastic rate; his or her body grows rapidly. Then the rate of growth slows until one is only just holding one's own against decline. And within these differing phases the pace of life alters; from the ever-impatient child living eagerly for each moment, to the stable adult steadily consolidating his position, seemingly inactive to the younger populace. Time is relative to each age and each personality. You are as young as you feel, you can also worry yourself into an early grave. It's a state of mind, your viewpoint, which matters.

The sage saw in every progress a decline in another area. He saw no reason why progress should continue forever. It could easily cease and become regression at any moment.

Having recognised the three types of change, the sage wondered what set the changes in motion. Observing the seasonal changes from summer to winter, he believed a change from light to dark force was responsible. Similarly, in spring the dark force waned and the light force gained in strength.

The Tao is one, a unity, but in order for it to manifest itself in this material world it branches into two primal powers of nature: the light and dark forces; the firm and the yielding. Like two main branches emanating from a tree trunk, these two forces appear to diverge, whereas in fact they complement each other. Their difference in placement creates potential but they do not combat each other. They are two differing aspects of one homogenous organism; in no way can they be considered a duality.

These two forces, or powers, rise and fall. Tension arises, causing constant regeneration of cyclic change. If the duality remained in watertight compartments the whole power of interaction would be lost. This is how the Tao turns the potential into reality. The polarity between activity and receptivity maintains tension. Each adjustment manifests itself as a change. So life is seen as the ongoing cyclic alternation of tension–change–tension, etc.

Having withdrawn himself into seclusion to cultivate these insights, the sage now had the problem of making his observations intelligible and transmittable. At this time there were no books. Traditions were passed on orally by word-of-mouth and the sage knew words never express thoughts completely. He also knew a later epoch could not fully understand a former age; another sort of generation gap but on a longer time scale. Chinese philosophy overcame these obstacles by the use of images combined with symbols. In this manner, each complex condition of change could be likened to an easily understandable natural situation. These ideas would then stimulate receptive minds across the barrier of time. For example, the concept of steady perseverance was shown by water running downhill. Although water is a weak and pliable fluid, when it meets an obstacle it gives, envelopes the obstruction and then passes on. It persists in its journey towards its goal, regardless of impediments.

What viewpoint did the sage arrive at?

This is broadly what he thought.

In the world confusion prevails, yet there is a system of order pervading this apparent chaos. It is *this* system we must attune ourselves to if we are to live in relative peace and harmony. And, as in all major religious and pragmatic philosophies, this means boring old moderation, integrity, humbleness and virtue. In Greece the colour of moderation was golden, in China that colour was yellow.

The early sages had a healthy view of destiny. They believed man had the ability to shape his own nature provided that his emotional life was correctly centred. By making subtle changes at the beginning of a course of action, the end result could be selected. Hence the sage could envisage the outcome of an event by understanding the consequences of each action at a given time. Correction had to be made early, before the situation grew too big to handle, or alter. On any journey fine trimming of direction can have an amplified effect upon the destination. All this fits in with tree theory quite nicely.

If a sage were to study an individual's tree of life, he should be able to ascertain future trends and say where, for example, encouragement should be given and support withdrawn.

I wonder what the Old Masters, the sages, would think of this dissertation. What did they think of conflict, politics and all the other things we have encountered?

Firstly, they would have said these ideas are not mine. There can be no authorship of ideas; all are generated by The Principle and merely await the appearance of a resonant mind to give the ideas substance, or to make it entertaining. And in order to do this, it is merely my task to stimulate that part which is inherently receptive in each individual reader.

The sage believed there could be no true battle between man and nature. Man was a part of nature, undifferentiated from it. Therefore, the sage did not preach a back-to-nature philosophy.

He could not. Man cannot return to what he is. Nevertheless, man was seen to be in a unique position. Left to itself the acorn would naturally follow its inherent Way. However, man now has the power to maintain or destroy the balance of nature itself. Hence he has become increasingly the keeper of that balance. Today (instead) he is becoming the greatest disturber.

Conflict between geometry and nature could not be considered (back then) as geometry was unknown at the time. The sage knew there was very little in nature that was rigid or straight. Even mountains appeared to change as mists drifted across the landscapes, creating a sense of perpetual flux. So he strongly disapproved of codes of law and creeds. These gave a false sense of security; one merely had to obey and all seemed well. But in the ever-changing situations of life, rigidity is death, for the right actions at one time can be inappropriate at another. Rigid ideas are considered a prison house of man's own creation: hard blocks of prejudices, forms and names. We in the West like everything labelled, confined behind the rigid bars of a mental prison. This is beneficial for exact science, but insufficiently fluid for life. Life is dynamic, supple, and ever-changing. Death is rigid and static.

In China, roundness symbolised movement, dynamism and the circle of the heavens. Squareness was of the Earth, static and passive.

I expect these ideas and images to be strange to the reader because of the environment in which we now live. In Britain your education probably did not encompass world religion, only Western religion. We are fortunate, in this country, to be able to consider other religious ideas and cosmological principles. In devout Muslim societies this would not be permitted. The sage realised this difficulty of narrowness of experience and expressed it as follows: you cannot speak to a well-frog of the ocean, you cannot speak to a mayfly about winter, you cannot speak of The Way to a scholar for his scope will be too restricted. Worldly knowledge is considered to be false intelligence, facts about facts which only lead to further confusion. We often mistake naming of things for the understanding of them. Witness a young child who sees the Moon for the first time. She feels the moonlight; then it is named for her and the mystery is gone. The average person prefers the seemingly safety of the logical world, with everything neatly labelled and pigeon-holed. Then nothing unexpected or upsetting should occur and one will not be confronted with the unusual. This is seen as preferable to having to change or adapt. This attitude, though understandable, is a static one and dams up the source of all wisdom, the wonder of the open mind.

What was the goal of the early sages? It was different from my goal of balanced all-round growth; nevertheless I feel they would have had great sympathy for this concept. Certainly, they rejected all extreme positions. Any exaggerated obstinacy leads to ruin. Wisdom, they maintained, consisted of holding oneself in the centre, neutral and indifferent. Before any action the sage examined the goal and chose the means to attain the end. Alternatively, he chose a course of non-action. This meant yielding to oncoming forces in such a way they were unable to harm him.

In Eastern metaphysics there is no clear-cut black-and-white attitude, as there is in Western logic. Nothing is absolute. There is always an alternative, a third element. This may be a reconciling notion, a compromise, or it may be an active decision to do nothing and let events run their course. Some things can be developed according to their nature, but it may be unwise to force them along. The former (passivity) is controlled sustained power; the latter (forcing) is quickly spent and dissipated; so this apparent passivity is more enduring than direct action.

The sage lived a life beyond the concepts of good and bad. He had no deficiencies full stop, therefore he faced no moral dilemmas and was untroubled by doctrines of sin. What we consider as sin he thought of as ignorance or stupidity which would bring automatic retribution out of the laws of nature. Each animal, plant or tree had its own nature, its own appropriate way of living, a destiny enacted in surroundings of continuous change. The mayfly, having conformed to its nature, is happy to die in the evening of the day it was born, having survived to its full life expectancy. The aim of all organisms is longevity; those with a better understanding of The Way can enhance this goal. Those who flout The Principle exist under unnecessary strain. Acts of folly contravene The Way and any violations are self-correcting, stressing the individual involved. In Western theology we are obliged to conform to the will of God, or else. In contrast, the sage fostered a natural desire to co-operate with the harmony of the Universe.

A sage was once asked how to make trees grow, for a garden retreat was of high importance, somewhere to sit quietly and contemplate. Japanese gardens of today still attempt to capture such tranquillity. The sage made his garden in the manner of nature. He did not have square lawns surrounded by angular fences. Instead he tried to blend areas of open space with a mixture of trees and stone. To a casual observer his garden could easily be mistaken for a creation of nature, which is exactly what he intended. Only the trained eye could see further into his designs.

Trees were highly placed in Chinese symbolism because of their diversity in unity. Also, when the tree came of age and bore fruit, each seed contained the one again. This presented a powerful image to the Chinese mind.

The sage knew the best trees were grown by avoiding trying to make them grow. He avoided scratching and shaking and firming them into the ground. All he provided was an initial good environment (good soil) then he left the tree alone to grow naturally. And he felt if the poor people were also left alone, instead of being badgered and advised from morning until night, they, too, might improve their lot.

Nowadays people are striving so hard to be different they all end up the same. You know: each desires a car, or double-glazing, then a video, or a deep-fat fryer; next we all buy microwaves, then we all buy compact disc players and so on. We live in an age where most people are anxious to educate their neighbours, leaving no time to educate themselves. I feel more self-culture and self-development is needed; so did the sage. Eastern philosophies were regarded as useless if they had no positive effect upon character. They aimed to produce the perfect man – the sage. Wholeness is required of the sage and attainment of this maturity comes only with the ability to accept and reconcile all opposites. Tension between opposites is now seen as a necessity to fuel actions; it is an impersonal requirement of any dynamic system. When one understands this conflict becomes something lesser; argumentation perhaps, or disagreement between two parts of a whole. This is far more favourable than enmity leading to a fight and a division of the whole into two separate entities. Disunity is no achievement.

Tree theory has a lot in common with the wisdom of the sages and has provided me with the solution to Linda's tangential question: what do I believe in? Answer: The Way which is mediated by the fundamental law of the Universe.

I have come to believe, from empirical observation and by learning, we live in a Universe where **asymmetry is latent failure**. I feel this is possibly *the* fundamental law of the Universe which, like Darwin's theory of evolution by natural selection, may not be expressible mathematically. This is The Way of the Universe. It is applicable from the sub-atomic particle level to the macroscopic world around us, then further out into intergalactic space.

If any system is unbalanced it will become unstable and degenerate to a more stable, less complex entity.

This universal law applies to machines and people and biospheres, and is relentless in its correction of imbalance. Later on I speculate on how the law is manifested and mediated in physical reality, but for now let us return to the sages. They seemingly took the concept of non-action too far, in fact to extremes (I would say), simply isolating themselves from the world to let it unfold as its predilections desired. I feel this was a step too far and not acceptable. Imagine neighbours who never tidied their gardens and let them run wild through non-action. An atheist may quip the resulting overgrown morass is what happens when you leave it to God, but tree theory does not condone this state of affairs. Action is the norm; non-action is a useful ploy in certain circumstances only.

Also there is this business of cycles of life and death.

Individual trees do not have cycles of life, death and rebirth. We can observe this and see it as a fact. A tree is either living or dead. Why should man be any different to the tree or any other species on Earth? The complex interaction of cells combining to make up the individual organism, which others recognise as me (my body), will one day cease to function. If so, why not believe in rebirth or a place called Heaven to lessen the worry of this imminent moment? Well, it is a desire, no matter how painful, to reason out the truth. The spirit which moves us to seek the truth is the truth that we seek. When the individual is not alive he, or she, is dead. That is the physical situation. But this is not the end of the story for the tree or anyone else. Although one's individual life is a one-off finite affair, it was brought about by a genetic arrangement and the individual has the capacity to leave a modified genetic legacy in his, or her, offspring which surpasses death. The selfish gene lives on and concurrently there is a parallel cultural inheritance which we each encounter, modify and pass on.

Times change and our culture has to move with them. Consequently traditions alter and we each teach the next generation slightly different customs...

Linda comes home from school one day near to Christmas and announces Father Christmas has a magic key to enter children's houses. Teacher said so! Who am I to argue with a teacher? I suppose all this central heating has led to a vast reduction of chimney pots in the yo-ho-ho size-range; so why not a magic key?

The individual dies and is dead, but his past presence can live on through kin and culture. This is very near to the truth of the matter, I feel. That's why I can still read about sages and you can digest this epic.

It is time to leave The Way for now. To go on labouring a point might make the open-minded reader contemptuous. It's a matter of timing.

One thing my studies and students of The Way understand is the immeasurable importance of timing. There is good timing and bad timing: a right time and a wrong time for each and every action. Saying, or doing, the right thing at the right time is one hundred per cent better than doing the right thing at the wrong time. This appears obvious, but how do you teach timing?

"Dad...Dad?"

"Yes, Linda."

We are sitting in a busy cottage hospital waiting room. Linda is about to have a hearing test today. It is a precautionary measure after a series of troublesome ear infections.

The waiting room is tastefully decorated with a three-dimensional collage made by children from a local school. The scene is a condensed panorama of a small town with shops and a market, a church with a cemetery, a hospital, cars and people, and sheep and cattle grazing on a backcloth of sweeping green baize hills.

"Dad ... Dad?"

"Yes. What?"

"Is that our town?"

"Well, sort of. It's not identical."

"Why?"

"Because ..." We're at it again. It's think-of-an-answer time. "it isn't big enough to fit our town."

Linda continues to look at all those pieces of polystyrene and milk bottle tops and hand drawings and cardboard which have been patiently glued together to make the glass-encased scene.

The montage does, in fact, tell a story from left to right. You are born in the hospital, baptised in church, grow up and spend all your money in the shops, drive a car and eat the sheep and cattle; then end up in the cemetery.

"D-Dad?"

"Yes, Owen?" Owen is waiting with us too as Tracy is elsewhere doing some vital shopping. Owen talks slower: he starts with a little stutter then continues fluidly once his linguistic engine is all fired-up. Time has moved on again, as it tends to do. Linda is now five years old and Owen three.

"D-Dad. Look a' dat funny man." Owen points down the corridor towards an unfortunate man with his arm held horizontal in a plaster cast. A sea of faces follows the direction of Owen's finger. I try to think of something to say as my face turns red. He could have said it quietly ...

"Dad ... Dad."

"Oh yes, Linda."

"Why have you got your best shoes on?" she asks loudly.

Everyone looks. I have to smile and when I look up from my shoes one or two other faces are smiling, too.

"D-Dad." It must be Owen's turn to embarrass me now.

"Yes."

"W-why has Mummy left us?"

"Shhh ..." I expect everyone to be frowning now, but a furtive glance finds our audience with even broader smiles etched on to their faces. "Shhh ... shopping. She's gone shopping."

"D-Dad." It's Owen again. I thought it was Linda's turn to make me cringe. "D-Dad, I wanna wee."

Oh gawd.

"Linda, will you stay here while I ..."

"No."

"But we won't be ..."

"Want to come with you."

Owen is bouncing up and down on my knee clutching his thingy through his trousers.

An elderly lady next in line comes to our rescue: "I'll keep your place," she says.

"Oh, would you? Thank you very much. Come on, you two." Off we go in search of a toilet: past the man with the funny arm ("Morning"), and ... Yes! Here's Tracy back from shopping! Good timing, girl. That's the way. That's *The* Way.

Evolution is testable because Darwin described a *mechanism* to drive it, survival of the fittest by natural selection. This elevates evolution to a scientific theory.

My fundamental law of the Universe "asymmetry is latent failure" remains just a theory. Yet over the last thirty years I have applied that theory and observed its inevitable consequences in machines, people and systems of organisation in general and have not found it wanting. In a machine one can quantify an imbalance by measuring physical parameters. Perhaps one component takes more than half the load current and is prone to failure as a result. In an unhealthy person, doctors try to correct the imbalances with drugs or intervention. Yet it is the hardest thing of all to try to correct a system of organisation obviously out of kilter. Those inside the

organisation prefer adding on bits here and there rather than re-organising the structure properly from the bottom up. These older organisations eventually decline. I perceive this as the future for first-past-the-post democracy. Its divisiveness will eventual be its undoing.

Around this time (thirty years ago) we set off on our usual Sunday morning journey to see my Mum and Dad. Linda and Owen were looking out of the windows in the back of the car when we passed by a mature oak tree with one bare desiccated branch sticking right out of the green canopy by about four feet.
"See that branch," I said. "It's going to fall off".
It had obviously outreached its supply of nutrients.
And so we waited, and every week as we went by that branch stubbornly remained. Well, it was solid oak.
Then one cold February morning about eleven years later it was gone.
Thus the 7th of February became *twig fell off the tree day*: a day to check how the balance of your life is going, a day for reflection. By now Linda and Owen were old enough to understand what I meant when I explained anything unbalanced failed. Yet only I really understood this facet in depth, which is why I now try to explain my findings to you in this work.

Latent means dormant, or not yet manifest. In terms of asymmetry it means although nothing may have happened yet, if the unbalance continues, eventually correction will occur. I propose and discuss the mechanism for my theory later.

Branch 19

Purpose

Nothing is perfect. Everything changes.

Since we began this journey a number of imperfect changes have taken place.

One day in September, when the grass has stopped growing, a very large bulldozer enters the field behind our garden: ominous. Next a lot of little tractors and unusual looking trailers arrive. Lorries unload mounds of grey gravel just inside the gate on the grass. Then big reels of blue tube are deposited by the limestone gravel piles. All this activity is in the field where I once collected freshly dug mole-soil and admired the old stooping oak tree by the stile set in the hawthorn hedgerow. This same field where one can feel the open space like claustrophobia in reverse, close to the small woodland copse we used to explore.

The next day in September I arrive home from work and, as usual, the children greet me by the gate at the top of the drive.

"Dad … Dad!" "D-Dad."

"Come and see what …"

"N-no, me tell 'im."

I am dragged down the drive to the rear garden.

"Look, Dad," says Linda, "see what the tractor-man has done."

"Yes-yes, look, Dad," says Owen in a dramatic tone of voice.

The scenery is different. The view has changed. No stile, no hedgerow and the old oak tree (poor old oak tree) is lying down on the grass with its white root-ball exposed – a gnarled bundle of fibrous roots unearthed and suffocating in the air like a fish out of water; a sad change.

As the days go by the huge bulldozer and its service vehicles pull yards and metres of blue tube into the ground. The furrows are back-filled with gravel then the land is levelled again. What used to be two homely fields now becomes one featureless expanse of criss-crossed mud.

The drainage team leave, then the local farmer arrives to till the topsoil and reseed his new prairie. The old oak tree is dragged to one side and left to decompose, but this is not the last we see of the old oak tree. A year later, after the farmer has had three good crops of grass from his new savannah, he starts to chop up the old oak tree with a chain saw. Firewood, I expect, but no; he is making fencing posts instead.

All around the inside perimeter of the field, oak fencing posts are pile-driven into the ground at regular intervals. Between the posts the farmer and an assistant stretch two perfectly horizontal lengths of taut barbed wire. I cannot see any reason for this and remain puzzled. There is already a ditch and hedgerow to contain the herd, why this extra barrier?

Another regression follows soon after. The wavy hedge which lined part of my route to the bus stop is hewn horizontal; mechanised; digitised. Its serene analogue curves are capped and the sides sloped, making it pyramid-like in cross section. And so the trend towards larger hedgeless fields and level horizontal hedges continues. But these tendencies are not irreversible. Neither is the system which fosters these actions.

We have been examining two systems of our own small planet. The signs are not good. While the pyramid predominates, the tree cannot flourish. The Earth's self-regenerating and regulating mechanisms are being exhausted and stressed by pollution of our own making. And it may well be, while we are all pondering upon the threat of some nuclear holocaust, a creeping imbalance of the ecosystem will slowly overwhelm us totally and unexpectedly. The air is no longer pure: bees smell fewer flowers through the traffic fumes.

[Remember, I wrote the above thirty years ago, before the media realised the Industrial Revolution caused the onset of global climate change, at a time when climatologists were expecting another ice age].

Our planet already possesses extremes, not just climatic imbalances, but economic, religious and political ones. The latter we have already discussed sufficiently.

Economic extremes are a necessary requirement for the capitalist, who venerates the pyramid and only aims to climb it and not change it. His actions enhance the affluence of the few by leeching the aspirations of the many. Insatiably he yearns for more gains, although his needs were satisfied a long time ago. Now he is fuelled by habitual lust. The figures for world debt are even more staggering than those I mentioned on nuclear weapons. The debtor countries owe one thousand, thousand, million US dollars. Crikey!

Religious divides seem, to me, to take on unnecessary and (in some cases) absurd proportions. Theologians locked in debate over their faiths behave no better than a school child arguing over whose big brother is the best fighter. Does it matter? Moral decay is ravaging society while the self-proclaimed custodians of morality launch another sectarian atrocity on a differing faith. Is it any wonder some people can no longer discern right from wrong?

Ecologically the planet is becoming a mess. We are given the message regularly on media reports. Yet realisation these disturbances to the biosphere are potentially serious is evident to only a powerless few. To the rest, it is either incomprehensible or a joke.

Here are some of the premier threats to the environment. Acid rain, nuclear fallout from accidents or otherwise, deforestation – especially around the Amazon basin, a hole has appeared in the ozone layer and carbon dioxide build-up may set off the greenhouse effect, heating up the planet.

The world is overpopulated. Rivers, once fresh water, are now polluted by chemicals, oceans are smeared with oil and ditches (ah yes, ditches) are filled with toxic nitrates drained from high production grasslands. That's why they put two lengths of geometric barbed wire on old oak tree posts on the inside of fields – to stop the cows drinking the polluted water. The residue of pesticides can now be measured in all creatures on Earth as bad farming practices disturb global areas untouched, as yet, by industrial pollutants. All these effects are known to be man-made and all are most definitely damaging to species. We are doing to the global Earth what our ancestors did to Easter Island – completely draining all its finite resources. And remember what happened to them.

Our planet is an anomaly in the solar system. The Earth's climatic system has maintained conditions suited to life for three thousand, million years. As far as we know, all other planets are lifeless; dead. In this long time span there have been many major disturbances to the ecosphere, yet life has always responded by adapting to the changes with great flexibility. And in doing so life alters the atmosphere for its own ends. Put another way, life maintains the condition for life. This is another example of cyclic logic, tension between two opposites yet complementary factors: life and Earth's accrued energy gradient (in the form of chemical mixtures and the Sun's radiant energy). The two forces rise and fall to create a planet with optimum potential to support life.

If we were to draw a political, economic, religious and ecological pyramid of the world and invert it, we would have a much disfigured, gnarled, twisted, unbalanced, unhealthy looking tree.

What can be done to correct this? Well, if the majority believe the Earth was made for man and have faith in that notion, there is nothing we can do but sit back and watch the planet destroy another unbalanced, unfit, species. This would be non-action in the extreme, because at some time during the species' decline, reality would dawn on all but a fanatical few. By then it may be too late for us to appreciate the Universe which made us was not made *for* us. We have no more right to be here than any other inappropriate species. We do not own this planet, but can be seen as leaseholders able to set our own expiry date. Indeed, pollution to one side for the moment, the larger land-based animals play only a small part in the Earth's self-regulating system. Micro-organisms along the continental shelves and in wetland areas play a far more vital role in maintaining homeostasis. In terms of overall life of the planet we are unimportant, so if we did have a nuclear skirmish which obliterated man, the bulk of life would continue largely undisturbed. Mankind's branch may amputate itself from the tree of evolution, but in time another species would evolve from some secondary shoot now lying dormant lower down the tree; just as we mammals came to the fore when the dinosaurs relinquished their supremacy.

Disharmony is potential doom. We cannot be blasé enough to assume everything will turn out fine for us or our little planet. We must aim to ensure everything needing attention is at least attempted. Mankind's future is entirely dependent upon our own efforts. We cannot expect guidance from the skies.

The Universe is non-ethical: it treats the good and bad with the same indifference. But this impartiality preys upon the weak and imbalanced, making them less likely to succeed than the fit. So if we, as individuals, are to flourish we must strive to be healthy and whole. This is the individual's prime purpose. The Universe has no omnipotent purpose out there waiting to be discovered. It is simply constrained to be balanced.

On Earth different beliefs pursue varied purposes. Since there is no universal purpose we no longer need to waste time arguing which one is right. The Christian purpose seems to be to worship God while alive, to attain everlasting life when dead.

What is the purpose of The Way?

The Way is not at all like Christianity or any other formalised religion. It is not taught formally, nor can it be. Each man and woman must find in themselves their own truths, then they are on the path to The Way. But the Tao reveals itself differently to each of us. Each would design his own unique coherent system if the Establishment and the media would leave them alone to do so.

We function upon ideas constructed through various phases. There may be likenesses, but each sees the world in a unique way. Sets of ideas which individuals adhere to may be similar, but no two people think exactly the same overall. Each is a complicated Universe in miniature; a complex balance and blend of forces swimming along in a sea of change. To each the purpose is different and can be so because no overriding purpose stands in their way. Yet the constraint is the same; dynamic balance is best maintained.

On our journey I have mentioned terms like the Golden Mean, balance, harmony, humbleness, moderation, and have implied they contain a certain merit and should be pursued. Well they should; but do not overdo it. I don't want you to become a bore and project an image of a slow-witted dolt; it's just I do not want you to become unbalanced, unstable, arrogant, dogmatic or rigid in your outlook either. Remember at all times you are individual and unique; you weren't made to be bland. Don't become mediocre, become a catalyst for moderation. Promote common sense at every opportunity.

Regrettably, I feel a lot of gifted people become plain because of the feeling of alienation rigid hierarchic order fosters. They abandon their own path along the way and instead find some form of contentment in pointing out the direction to selective travellers who wander by. I urge these people to rise up from their comfort zone and get back on the path again. The assertive will not allow the meek to inherit the Earth by default.

So promote your own canopy committee. Demand a Christmas tree with roots on. Don't settle for second best. Submerging one's individuality in totality will not benefit the whole, for the dynamic interplay necessary in a healthy society needs every member to make some positive assertions. This I have tried to do in this work.

Individually we can affect society. If each individual decided overnight to give up, say, sugar, the sugar industry would collapse and die. If each individual decided not to succumb to greed capitalism would falter. If our collective unconscious minds eliminated fear and hatred, wars would end. This is the power and the predicament of the individual.

The individual can influence events. I urge readers to take on your own challenges and responsibilities; discharge them to conform to your own way, not that of an impersonal pyramid system. And if you feel like printing the pages of a book in a different, non-geometric style, do it. Meanwhile, shun the pyramid-people of this world; ignore them: their time has passed.

I have constructed a theory
to fit my life. It will not fit yours,
but no matter. I expect you to be different;
I expect you to disagree on many points. My theory
is designed for adaption. The whole book should be held,
like a manuscript, in a ring binder, where elements can be
added, or subtracted, if the need arises. I give you an adaptable
tree theory which you can adopt and nurture to your own
requirements. Some will claim this is a cheat,
but it most definitely is not. The cheats are
the ones selling you dogma.
They preach to you
unalterable tenets
fixed for all time,
which rarely, if ever, exist.

The journey is not over,
though soon we must part company
to pursue our own destinies. But there is
no destiny, except that which we manufacture
for ourselves, unaided by a maker. But there is
no maker, except for those in doctrines maintained
by theological experts. But there are no experts, only
deterministic academics. And determinism is flawed, we
cannot foretell our purpose, for no omnipotent purpose
exists to guide our ethics; whilst morality is left for
the media to distort to excess and pontificate over in
the name of truth. Yet even truth
is relative. Indeed, we
ourselves are relative,
relatively unimportant
in terms of global life.

All this sounds somewhat negative, though it is not. It is a necessary part of the searching process to rediscover the peripheral artefacts of culture which may have been useful in the past, but are now becoming out of fashion, like co-operation.

When I began, Linda was a mere seven months old, now she is seven years old. Yet this passage of time can now be viewed from a more complete perspective, for Linda and I are two entities at differing points on the semicircle of life. I am nearing my zenith, while she is rapidly ascending into the world. We are members of a species which is nearing its zenith, too, spinning along on the only planet which sustains sentient life in a solar system of finite duration. In ten eons the Sun's nuclear globe will expand, making the Earth arid and lifeless. Our nice yellow furnace will then transmute into a Red Giant whose outer corona will contain the existing orbital radius of the Earth. No more Earth: then a while later our galaxy may be sucked into a black hole. Then all the black holes may conglomerate into some egg-like lump of matter where ultimate symmetry will once again be attained and time will stop. The whole Universe will again be at one point, at one time. This is how it was in the beginning; pure symmetry with no time. It was like this before the Universe exploded into asymmetry and time.

[Now it has been discovered the rate of expansion of the Universe is increasing, so now no Big Crunch. And predictions of a return to an ice age have diminished, too – all discovered in thirty years].

Hence we can see little cycles of activity in different phases of development inside larger cycles of activity at different points of evolution, revolving inside the bigger and ultimate cosmic cycle of change. We are a very small cog in the wheel of the Universe and unless we tune into the harmony of its Way we will be discarded as individuals, as a species and as a planet. Man-made pyramid systems function in a blinkered, insular manner, which are in discord with the overriding scheme of things. They are depleting and destroying the bounty of reserves and life with which our planet is imbued.

How best to teach all this to a young girl?

Surprisingly enough, the subject matter is ideal for the young. It is when time has hardened one's sensitivity that learning ceases. There is no fixed lower age limit for students of The Way, only an upper limit when the student becomes not-young-at-heart. However there are practical considerations. Linda will not be able to read this book with understanding until a certain age of development, probably in her teens. Then she can make up her own mind about the two systems of a small planet her Dad has encountered, before she becomes set in her ways.

The perspective of living on a planet sailing amongst the stars is clear to Linda even now.

When I was a lad of the same age the notion was harder to appreciate. Since then we have seen live television pictures of men on our Moon, accompanied by glorious scenes looking back "down" on to the blue marbled globe of Earth. It is even possible to simulate spaceflight with a variety of new computer games.

Meanwhile, back on Earth, daily happenings often throw up events which seem strange to a child. Linda is always keen to pursue questions based around some anomaly in her young logic. Often her logic is correct and the anomaly stems from adult acceptance of the logically absurd based upon some abstract system of organisation. Sometimes it makes me wonder who is teaching who. Perhaps, as in tree systems, it is the two-way exchange between us which will bear the most fruit. Then when she is at her zenith and I have fallen down below the horizon she will recall the absurdities of the twentieth century and act to eradicate them.

There is one looming danger remaining which I wish to explain. Having resolved many issues in this work to my own humble satisfaction, please do not think this places me beyond persuasion. Yet once a coherent systems philosophy is constructed, one does tend to rigidly adhere to it. This gradually exposes one's position until someone else forms his own concept of the theory and hangs this different version around the originator's neck like a stone. I can see no method of avoiding this impasse except to try to remain fiercely humble and loudly adaptable. And, in any case, my systems theory is imbued with flexibility.

I have found simple tree theory very effective in balancing my life. My financial branch is stable at present, just the right size to cope with the rest of my outlets, but I know it will be stressed shortly, so I plan for some economies in advance. My family branch I know not to expand! Three would be a crowd and upset the whole balance. Our little family unit is right for the rest of the tree.

At work my career fork expands again in a new interesting direction, partly due to the interest I show: note the circular logic. There is even talk of "flexibility" and communication meetings begin again in a slightly different guise. Change is in the air.

Linda goes to school, is good at her work and has changed from a polite little girl into a lively rascal. Owen learns naughtyish things from Big Sister and starts off as a rascal who will probably go quieter when in school; for every gain there is a loss.

Tracy starts to teach again, not in secondary school, but as the head of the local playschool. This does not pay very much but it is very fulfilling. Even I enjoy getting involved in all sorts of playschool activities: hat-making at Christmas, toy repairs, summer carnival fund-raising events and helping out at jumble sales, etc. It is hard work, but not *all* hard work; it is good fun, too. I even make the Christmas post box for the children to post their letters to Santa.

As I close this story, to selfishly give myself more time for all-round expansion, Christmas is approaching once again. The super-powers are talking about arms reductions, so perhaps there will indeed be peace on Earth one day. However, Linda is starting to ask some awkward questions about Father Christmas; like how does Father Christmas know what everyone wants? And, why have some presents come from Auntie Alice and Uncle Harry? How did they get their presents to the North Pole for him to deliver? Also, why – why, Dad – can't we go up into the attic?

"It's too cold in winter to play up there," I reply.

"You went up there," announces Owen.

"When?"

"Last night, Dad." Accusing tone of voice, laced with intent. "I'eard you." So there. (I thought he was asleep.)

I suppose a seeker of truth, as I sometimes claim to be, should have a guilt complex over this Father Christmas issue, but I don't. I believed in him when I was five, thought there was something fishy going on around eight years old (especially when my Dad kept popping up into the attic) but now I am a firm believer once again. It helps if you tend to be young at heart. Granted, there is a lot I do not believe in, but I do believe in Santa C. He is such a powerful personification of all that is good and cheerful no one has qualms about him; no one really objects. People may complain about other concepts, but there are no Santa-agnostics. Nobody walks around with "Santa is a fraud" placards, do they? There is no pressure group lobbying MPs for a Campaign to Reveal All about Synthetic Santa (CRASS?). No, the man is watertight: invincible. And tomorrow we are going to visit him in his grotto in a department store in the big city.

"Dad, where are we going?" asks Owen. All traces of a stammer are gone now; now his big sister has been instructed to let him get a word in edgeways now and then.

"To see Father Christmas, Owen," I reply, somewhat exasperated, having explained all this before at least twice. "Have you both got your letters for him?"

"Yeees, Dad," says Linda.

"Where's mine?" says Owen.

"In my handbag," Tracy replies.

Traffic is heavy. We meet the queue into town at a roundabout on the city's edge and this is at ten to nine in the morning, before the shops have even opened. Fortunately we locals know a few detours around the backstreets and manage to clock-in at the multi-storey car park at one minute past nine.

Tracy and I have pre-arranged our expedition. I will take the children to see Santa while she buys the toys he is supposed to bring. We will meet by the grotto later. If you go Christmas shopping without a plan, you are doomed.

Into the store we go; through the racks of clothes: pullovers, dresses, coats, shirts and skirts. They all swish around the heads of Owen and Linda while Dad steers from above with subtle tugs on their hands. Now we reach a staircase festooned with white teddy bears; bears on the banister, bears on the shelves, bears hanging down the walls on tinsel and string, bears everywhere. There's a competition to guess how many teddy bears are in the store and first prize is... a white teddy bear. Slowly we ascend the stairs, the children clutching their gaily coloured envelopes, hand crayoned.

Now we are in the bridal suite. Dad's taken a wrong turn. On we go through the warren of alcove rooms: light fittings: bedding linen...

"Dad? Where is he?"

"We'll find him soon." I hope.

"We're lost," states Owen.

In the distance over the display of perfumed gift packs a whirling helicopter on a string cuts out a circular path beneath the winking eye of a watchful security camera. Aha! The toy department at last. Yes, and there is Santa Claus himself!

"Hello, Father Christmas," I say boldly as the gentleman in question ventures forth from his allotted green alcove and strides forward beside the obligatory Christmas tree.

"Hello, children," he says in a faintly Widnes accent. He stoops down to Owen's level, "It's nice to see you. And what's your name, young fellow?"

Owen goes all shy. He squirms down in his duffel coat like a tortoise retracting into its shell.

DAD: (Prompting) "Speak to Father Christmas."
OWEN: (Barely audible) "Owen."
SANTA: "And how old are you, young Owen?"
OWEN: (Whispering still) "Four."
SANTA: "And what do you want me to bring this Christmas?"
OWEN: (Extends out of his duffel coat. Audible now) "A Lego petrol station."
SANTA: (Overtly jolly) "Good! Be a good boy and I'll bring a few surprises, too." (Father Christmas tickles Owen under his

now visible chin then turns to Linda). "And what about you? What's your name?"

Linda bites her lip and sways on the spot, speechless and smiling. I prompt her to speak up or her sack will be empty.

Marvellous isn't it? All year around I'm reminding her not to speak to strangers and now I'm provoking her into talking to an old chap with a white beard, a red costume with white frills and a close association with elves and fairies. But Father Christmas can do no wrong and eventually Linda confesses she would like an ice-cream maker and Father Christmas says (if she is good, mind) he will bring a few surprises as well.

"Now post your letters in the box, children," says Santa. "Is there an address inside, Daddy?"

"No, Father Christmas," I reply, somewhat puzzled.

"Better write it on the envelopes then," he says winking at me, then (quieter to one side), "they might get a reply."

"Fine," I say, feeling in my pockets for a pen. We walk away and two more parents try to coax a little girl to talk to this nearly complete stranger dressed in red from Widnes.

Tracy has a pen in her handbag and a plastic carrier bag from the Early Learning Centre which now contains a few surprises, too. I fill in the address and we walk back to Santa's post box. There's quite a queue now and I overhear one woman say: "Let's go to Owen Owen's; there's a Father Christmas there as well." What blasphemy. Surely there is only *one* Father Christmas?

After a brief shopping trip amongst hoardes of people we head homeward. We visit my parents on the way for my Dad's birthday party, then return home to settle down for the night. All this activity has forestalled a few questions Linda and Owen store up for supper time.

"Dad?"

"Yes, Linda."

"How did Father Christmas get to the toy store?"

Tracy promptly replies, "On his sleigh, of course."

"Well," continues Linda, "where was it?"

"On the roof," Tracy says, "with the reindeer."

"Isn't," says Owen.

We all turn to look at him with who-pulled-your-wire looks on our faces. "Why isn't the sleigh on the roof?"

"'cause there's no snow," Owen concludes. So there.

Oh, I see. Owen thinks you need snow for the sleigh to glide over the rooftops.

"Silly," says Linda, disdainfully, "he flies through the sky. He doesn't need snow, does he?"

"Does"

"Doesn't."

There is a short pause.

"Doesn't need a reindeer, does he, Dad?" Owen queries.

We give him that perplexed look again.

"'cause," he continues, "the sleigh is pulled by a car. Isn't it, Dad? He came up our street last year."

Ah! Owen is referring to the visit by the local Round Table and their festive float for charity. When Father Christmas wore glasses and looked like the local Liberal councillor.

I open my mouth to say something, which I haven't worked out yet, but feel saying something will be better than saying nothing at all, when Linda (thankfully) interjects ...

"Anyway, I'm going to wait up all night and see him," she declares. "I'm going to look out of the window and see him in the sky."

Looks like Father Christmas is in for a late night this year.

And so, dear reader, in keeping with the Christmas tradition, I must wish you a happy and successful future. I know this is futile, since success now lies, as it always did, largely within your own control. Nevertheless, my futile wish is a sincere one. Your Father Christmas is out there somewhere; your all-powerful personalised concept is waiting for you to uncover, as I have uncovered mine.

The process of discovery takes place on two levels: first, the journey is through the real world, as if riding along inside a passenger coach on a railway track, looking out of the windows and seeing events unfold. Since my journey began, anxieties which I felt over some issues have been physically attended to; from late transport to nuclear peace on Earth, good progress has been made. This fills me with hope, as one whose wishes have been fashioned into reality.

Second, there has been a concurrent inner journey, within the vehicle (as it were) altering my previously conceived assumptions. I see this change as a progression from a lower to a higher order of complexity. I no longer expect or demand things to be perfect, correct or right, and resolutely believe in the predominance of disorder which is seen, empirically, to comply with universal tenets. I know no two days are (or will be) the same and regard normality as the exception in an environment full of fluctuation. This makes me less disturbed by routine chaos. I accept if one is good, reward does not automatically follow. Reward requires persistence: quiet, patient persistence. I overview most situations from a wider perspective now, then fill in the detail later when it is time to be attended to. But these newly-acquired characteristics (I will call type two person) are not always appreciated by others.

Some are disturbed by my now seemingly emotionless acceptance of unexpected deviations from their comprehension of strict normality. These are people of faith who believe that things should be right first time, regardless of circumstances. They possess a rigid, defined, limited and logically inconsistent myopic outlook and cannot envisage or behold a big concept, or formulate original plans and ideas. They are traditionalist, conservative, preservative and think if one is good things will turn out fine. In vain they try to foresee minute detail of undisclosed forthcoming events and when failure sets in stress, tension and anger soon follow. The anger is then focussed on the calm and quiet ones, who seem not to care about such details and (in fact) they don't care until, crucially, the time is right to act. In the long run, it's being effective that counts.

The quiet type two person treats the frenetic type one person of the last paragraph, not with anger, but with the same sympathetic understanding one would have for a lame horse.

I am now more conscious than ever before of the once-in-a-lifetime nature of everyday experience and have learned to openly appreciate the simple pleasures of life. Each event in the journey is to be savoured. We are not spectators of this journey, but participants in it. To understand the venture we must imbue our perceptual inputs (our sense data) with some meaning. Each individual creates his own Gestalt pattern, a world canvas of his own making. But if we step back from the canvas, all the patterns must blend into the cosmic order. Each event is once in a lifetime. But more than this, each lifetime is once in a Universe.

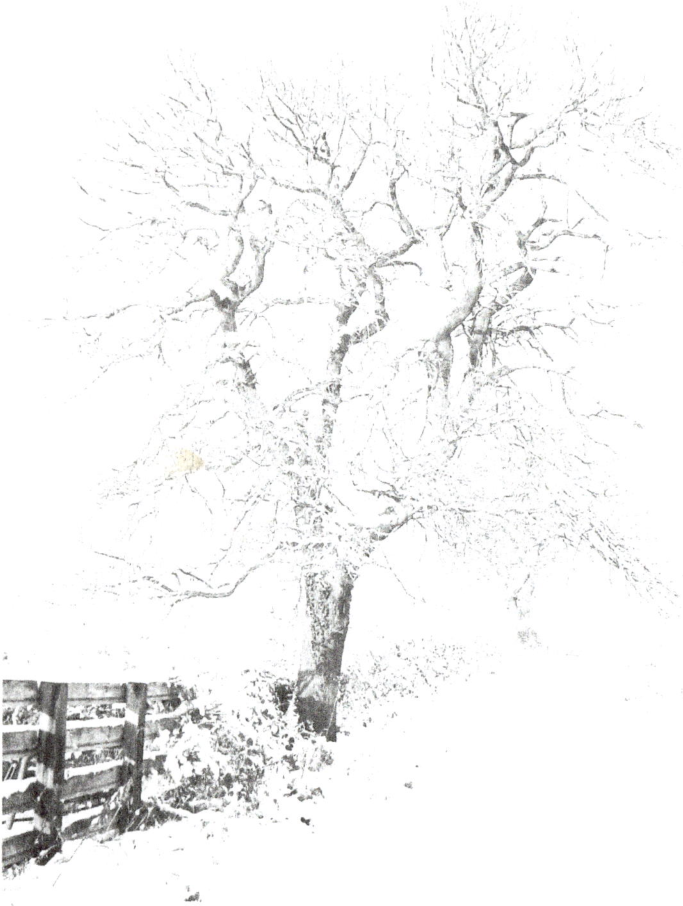

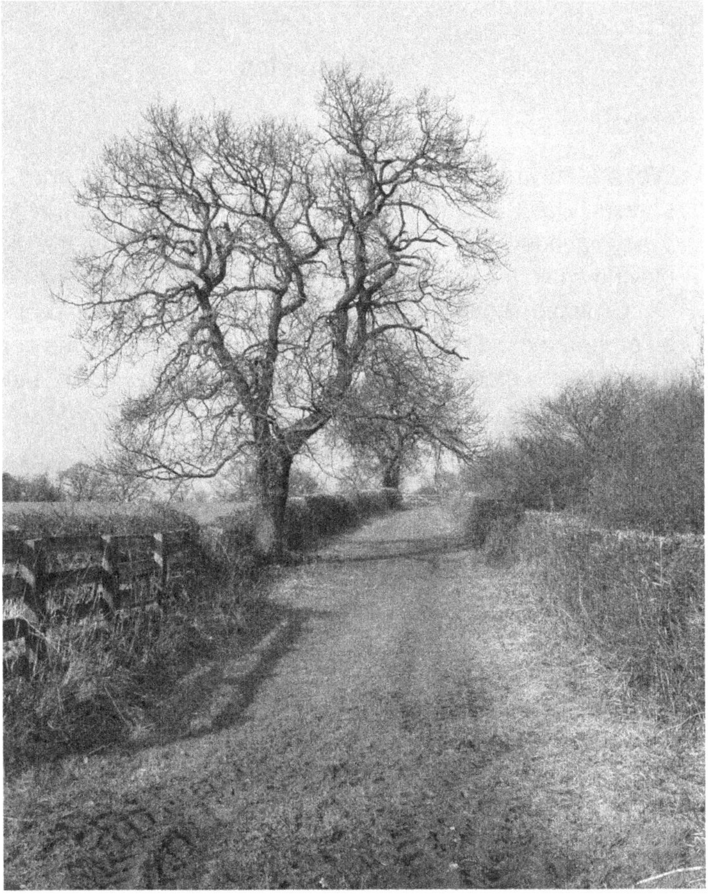

Epilogue

Metaphysics

We are having a family reunion. Owen is home from his university lecturing post and Linda has her children in tow. Tracy is busy with the buffet and I'm sitting on the floor playing snakes and ladders.

"Grandad, how old are you?" asks Linda's four-year-old elder son, reminding me of the time Linda asked me the same question all those years ago. But this time I am better prepared.

"Sixty-two," I reply.

There is a pause, then he says, "That's long," and we all laugh.

It has been a long time since I started looking for The Way, found an answer that pleased me, and lived my life in accordance with my findings. My tree theory is a secular thesis that can apply to individuals, organisational structures, machines and even electoral systems, as its basis (asymmetry is latent failure) is entwined in the fabric of the Universe itself.

But now this crazy old Grandad simply doesn't see the world like anyone else upon it. Not only does he no longer believe in time, he even questions if matter is all it seems.

Our friend Aristotle coined the term "metaphysics". He had completed his treatise on physics, then started his next work which he logically named "after physics". Now metaphysics means a branch of philosophy dealing with first principles, like time, space, substance and identity.

In this epilogue I try to explain how I think about time. Then I attempt to explain why asymmetry manifests itself as latent failure. I even have my own thoughts on particles and cosmology.

But this won't be an easy task, though it might help a lot if I say, right from the start, all this is pure speculation on my part and has no scientific basis. This is pure metaphysics, which is the reason I have extracted it from the main narrative of my thesis, within which it never comfortably sat. Yet I did not want to leave these speculative ideas unrecorded.

Remember the story so far, the Standard Model of the Universe is robust, yet the quantum world cannot be reconciled with gravity. Most of the Universe is dark energy: revised figures give dark energy at sixty-eight per cent and dark matter at twenty-seven per cent. Ordinary matter makes up the remaining five per cent of the Universe, the stars, the elements, you and me.

Also, about nineteen fundamental constants of nature (like the force of gravity) have to be deduced experimentally as they cannot be calculated by known formula.

The Universe is 13,800,000,000 (13.8 billion) years old. No one knows what happened at the instant of the Big Bang and trying to picture it in one's mind is all but impossible; but we are going to try anyway.

In the beginning there was almost nothing.

Nothing is very hard to picture, but I do it like this. First, I imagine the black full stop at the end of this sentence. Then I imagine a black page with the full stop on it, but now we can't see it, though we know it's there. The full stop is now incredibly small, around 10^{-30} centimetres in diameter. Then the fundamental law of the Universe is imposed upon this minute piece of nothing: a latent asymmetry spontaneously occurs. Vast, unimaginable amounts of energy pour in, creating a tiny seething expanding sphere. Space is created in which the energy is contained.

The Standard Model has four forces: gravity, strong nuclear, weak nuclear and electromagnetic, but at this time (just 10^{-43} seconds after the Big Bang) all these forces are united. The temperature is 10^{32} degrees Kelvin; it's hot.

Here I speculate on what happened next.

The spherical very early Universe is resonating acoustically. Have you ever seen a taut drum skin sprinkled with sand and a loud speaker forcing acoustic waves into the skin from below? Resonant patterns, peaks and troughs, are formed. Some areas contain sand; others contain no sand, just the outline of the resonant nodes in the skin. (This reverberation inside the early cosmos will later become visible, giving vast superclusters of galaxies a characteristic scale.)

Our early spherical Universe is also expanding rapidly in size, very rapidly (!) by 10^{30} to 10^{100} times in 10^{-32} of a second. This is superluminar (faster than the speed of light) expansion of space, known as inflation. At the end of inflation the diameter of the Universe has increased from the size of an atom to ten centimetres, about the size of a tennis ball.

So we have these two factors: acoustic resonance of space (in my metaphysics) and rapid expansion. As space expands it cools and different resonant modes occur, set by the available energy and sphere diameter. Expansion and cooling "traps" some resonances into the fabric of space. Each time this happens one of the fundamental forces is condensed out of the seething sea of energy and a grain size for that force is embedded into the fabric of space. In the standard model these are called "fields". Thus geometries are trapped within space. What form these take, I do not know. Perhaps it is the geometry known as E8. I imagine space now having a complex foam-like granular structure, a bit like a sponge, or perhaps grainy at the smallest scale as in the theory of Loop Quantum Gravity. This may give rise to the dark matter and the relaxation of the geometric grain of space may be the dark energy.

So let us just recap where we are. We are less than 10^{-32} seconds after the Big Bang. The temperature has fallen from 10^{32} to 10^{27} and pure radiant energy dominates inside our newly-created Universe. We are ready for the nineteen constants to be set and particles to form in the midst of the intense radiation storm.

The Universe must remain balanced; any asymmetries fail. This applies to the forces, the constants and the particles now starting to form from pure energy. In the Standard Model we have twelve matter particles and five force carriers, but at this time there are many, many more combinations being spontaneously created, "tested" and failing due to their asymmetry. Anti-matter is also being formed and annihilated though there is a slight bias in favour of matter: one part in around one billion. Thus, the Universe tunes itself. The only conditions which endure are those fitting within the self-set acoustically determined geometries of space.

We are still less than one second after the Big Bang. The Universe has very high symmetry and order; its entropy is therefore low. (Entropy is a measure of disorder and at this time there is very little disorder). The Universe continues to use asymmetry to fine-tune itself so today a deviation in the constants by one part in 10^{120} would make it unviable: it is incredibly well balanced. Today the energy in matter balances the negative energy in the gravitational field, making the total energy of the Universe zero. Also, electrons exactly balance proton numbers so overall the Universe has zero charge. And all this is brought about by the application of the fundamental law within the first second of the Universe's creation: asymmetry is latent failure.

One day (13.8 billion years later) I'm pushing a trolley around a supermarket, but I'm not thinking about food. Tracy is doing all that up ahead. No, I'm wondering how to explain the Universe to a shelf-stacker, one of the "Fruit and Veg Team".

Along the shelves are racks of pears, beef tomatoes and avocados, and each specific fruit has a box with cavities in it shaped to take that fruit. Thus the pears have rows of trays with pear-shaped cavities in the bottom of the boxes which protect the fruit in transit. Likewise, beef tomatoes have trays with large round cavities in the bottom of their boxes. Pears are the wrong shape for the beef tomato boxes and vice versa. If the fruit were packed in the wrong trays, by the time they got to the store, via articulated lorry, they'd be all bruised and damaged and in bits; unable to be sold and therefore asymmetric failures.

Meanwhile, back in the early Universe (still at 10^{-27} seconds), quarks and electrons are starting to form, but what size shall they take? The Universe doesn't have trays, but it does have (in my metaphysics) geometric acoustic cavities of space capable of holding specific energy values in each field. Some cavities are just the right size for electrons, others the right size for quarks.

Fractions of a second later the quarks are being pulled together by the strong nuclear force recently condensed as the Universe still expands and cools. Protons and neutrons are created which will soon form the nuclei of the elements hydrogen and helium. Within just three minutes all the ordinary matter is created and the Universe has a density ten times that of water. Now its diameter is several light years.

Now we have to wait around three hundred thousand years before the Universe is cool enough for the nuclei to capture electrons to form the elements hydrogen (seventy-five per cent) and helium (twenty-five per cent). At this point the Universe becomes transparent and changes from an undifferentiated soup of radiation to a matter-dominated universe where filaments of the elements can collapse under gravity to form the first stars and then galaxies.

Our Sun is (possibly) a third-generation star, formed from a gas cloud containing heavier elements of prior supernovas. The elements in our bodies also come from the collapsing gas cloud that coalesced some 4.6 billion years ago, creating our rock – planet Earth.

For years I pondered on how the Universe applied correction to asymmetric systems, be they machines, trees or people. Then recently it dawned on me it was all down to energy. If you have an asymmetry and you want it to endure then you have to feed it with a steady supply of energy for it to survive. If you don't, the system in question will revert to a more stable, less energetic state.

So for evolution by natural selection the mechanism is the survival of the fittest. For everything, asymmetry by latent failure is due to *insufficient sustainable energy*.

A badly designed machine can probably run for a long time if you keep feeding it sufficient power to sustain its asymmetries; but eventually someone will think "why don't we do such and such" and hopefully the asymmetry will be corrected before it spontaneously fails catastrophically.

Likewise governments can use the resources of the citizens to support asymmetric private ventures with no social and moral conscience, like banks. This is due (in the main) to the failed first-past-the-post electoral system that has disconnected the people from the power. Once the funding runs out the stresses will become greater, the asymmetry will be more self-evident and new ideas will be looked at in a more favourable light. This will take time.

So what about time?

I don't believe time exists. By that I mean time is not a substance; it doesn't flow through us, or have an arrow. The only arrow we are fighting is disorder, our old friend entropy walks in our shadow. Earlier I wrote "time moves on around us and we age regardless of our indecision." But now I would not see it like that and would write "matter moves around us and we age regardless..." Things move, not time.

I now think of time as a man-made and very convenient way of measuring the displacement of things. Yes, we measure time by the displacement of things; let's start with a day. We call a day the time it takes the Earth to rotate once on its axis; a year the time it takes the Earth to rotate once through the seasons on its orbit. Clocks measure vibrations of crystals and movement of cogs. None of them measures time directly because ... it doesn't exist. They all measure some form of displacement. Is there a clock in view, dear reader? If so, time is not flowing through it, it is being moved by some energy source. Take the battery out and it stops.

Not far from where I live is a canal boat museum where they have a row of terraced boatman cottages from the nineteen-fifties. The cottages have been preserved just as they were sixty years ago. Nothing has changed and nothing has moved. There's been no displacement and when I visit one room (in particular) it reminds me of my grandmother's house with the rectangular tiled fireplace and wooden furniture just like I remember it. Nothing has been changed and time has stopped in that cottage.

But, I hear you cry, that's crazy. If I say I'm going to meet you tomorrow at nine o'clock in the morning, time must exist. But what are we really saying here? Tomorrow, one axial rotation later, we will meet at a subdivision of the axial rotation. We call it nine o'clock, but it could equally be one hundred and thirty-five degrees past midnight, an angle, though this is not as convenient as the expression of displacement as expressed on a watch (which I am more than happy with).

Our biggest problem is entropy. It degrades everything. Our only defence is to try to eliminate the damaging asymmetries continuous change inevitably creates. One of my sayings is: "I maintain stability by planning for change." It's something to think about...

This only leaves me with my last novel idea to explain.

Physicists have been trying for about fifty years now to reconcile the quantum world of particles with the force of gravity and space-time. Now things are worse and all their theories (though powerful) are only expedient to the five per cent of matter we can see and test – the remaining ninety-five per cent of the Universe being (as I write) unknown. Perhaps when I update this book in thirty years it will all be sorted out.

Now for my second crazy idea: I don't believe in particles.

You see, a particle isn't a point of matter like a billiard ball or, say, a marble. A particle is not a solid object at all and is best described by a wave function. The particle's wave function describes the arena where the particle resides: it could be here or there, with this or that momentum or position, but all we can say in the quantum world is the particle is somewhere as expressed by its wave function.

In my metaphysical cosmology there are no particles, there is only trapped energy. When I look around at tables and chairs, trees and seas I see energy temporarily trapped in pockets condensed into the fabric of space created less than one second after the Big Bang. These pockets are like the fruit trays in the supermarket, they can only contain specific quanta. The reason all electrons are the same is due to their *containment* only permitting specific energy to reside within. The pockets are reminiscent of wave functions: the energy is inside the pocket, not at a point, but spread out within the cavity. As the Taoist sage says, the usefulness of a cup is in the cavity.

So what does reality consist of?

It is the dynamic placement of energy within a definably geometry selected by the law of asymmetry.

No time and no particles. It's heresy!

Neat, though, don't you think?

So finally, is there life anywhere else in the Universe? Sheer probability seems to make it almost certain. Billions of stars, billions of galaxies and planets being discovered all the time – one or two "Earth-like".

But I believe there is no other planet like Earth. Sorry. I know the numbers are large, but the probabilities are just too small to create anything like this rock we're on.

I've been trying to explain this to two physicist friends as we sit in a riverside pub. The tide is coming in and the river flows the wrong way – upstream. This is controlled by the Moon: how many planets have had a single moon for four billion years and seas, life and constant oxygen for two billion years, with meteor strikes to wipe out reptilian competition? How many planets have an iron core which generates a magnetic field to shield their atmosphere from the solar wind of their own star – on Earth evident as the Northern Lights, in our hemisphere?

Sorry, but the more I think about the chance events and asymmetries that have led to we three sitting by a riverside, the more I believe this is a one-off. But then our mythical Galum will also inhabit his own unique environment and may feel the same about his world. It probably won't rotate once in a day, or have seasons, and he won't measure displacement in seconds. Sometimes (when I watch the news) I wonder if, rather than being the pinnacle of evolution, we are simply a remnant born of past failures. If we do not find sufficient sustainable energy, the future will not be infinite.

No, as far as humanity is concerned this is it.

If we don't manage this planet successfully there will be no salvation for us anywhere else in the Universe.

This is a one-off.

Lightning Source UK Ltd.
Milton Keynes UK
UKOW06f1225180515

251752UK00014B/340/P